Wilson's Yakushima　Memories of the Past

ウィルソンの屋久島

100年の記憶の旅路

Tomoko Furui

古居　智子

序文

　荒川登山口から安房森林軌道を辿って約3時間、標高1030mの地点に大木の記憶を宿した巨大な切株がある。縄文杉登山道の途中にあり、また空洞となった中から見上げた空がハート形に縁どられていることもあって、屋久島の人気の観光スポットになっている。

　英国人植物学者のアーネスト・ヘンリー・ウィルソンが屋久島の存在を初めて世界に知らしめたということで、この切株は『ウィルソン株』と呼ばれ親しまれているが、詳しい背景について日本語で書かれた文献はほとんどない。

　ウィルソンが屋久島に上陸したのは、今からちょうど100年前の1914年（大正3年）2月のことだった。いったい、彼はどういう人生を送った人で、何の目的でこの南の離島までやってきて、どのような感想を抱いて帰ったのだろうか。

　その疑問に取りつかれた私は、2年前から主に英文資料にあたりながら地元の雑誌にウィルソンの伝記を書き始めた。そして、ハーバード大学アーノルド植物園資料庫に彼が屋久島で撮った写真が保存されていることを知るに至った。

　写真のガラスプレート原板は全部で57枚あった。山の奥深くに林立する巨木や、みずみずしい苔に覆われた花崗岩、傾斜を駆け下りる清らかな水の流れ、そして牧歌的な里の風景。そこには、まぎれもなく100年前の島の素顔が映し出されていた。

　夢中になって写真を眺めていたある日、風景の中に人の姿が写っていることに気がついた。巨大な樹木の隣で、あまりに小さく頼りなげに見える人物像を拡大しながら丁寧に見ていくと、そこに明らかに地元の青年たちと思われる着物姿で直立する生真面目な顔があった。彼らはその日その時、どのような思いでカメラの前に立ったのだろうか。

　写真のコピーを手に、登山口近くの集落のお年寄りの家々を回るうちに、少しずつ情報が得られ、ひとりまたひとりとウィルソンの写真におさまっていた3人の青年の名前と顔を特定することができた。そして、

Portrait of Ernest Henry Wilson, 1921.
ウィルソンのポートレート 1921年撮影

それぞれがウィルソンと過ごした短くも濃密な時間の記憶を三人三様のかたちで残していることを知った。その過程は、まさに秘密のベールが一枚一枚はがされていくような興奮と喜びに満ちた瞬間の連続であった。

　奇しくも今年は屋久島世界自然遺産登録20周年、さらに来年2月はウィルソン来島100周年にあたる。このタイミングを待っていたかのように長い眠りからよみがえり姿を現した彼らは、私たちに大切な何かを語りかけてくれているような気がしてならない。

　写真と文章でつづったこの一冊が、ウィルソンが屋久島に遺したメッセージの重みを今一度、受けとめるきっかけとなれば幸いである。

2013年11月

屋久島世界自然遺産登録20周年を記念して
古居　智子

Preface

On the island of Yakushima one thousand thirty meters above sea level, stands an enormous stump holding the memory of when it was a great tree. It takes a three-hour hike from the Arakawa trailhead to the stump, which is located along a trail that follows an old logging railroad to the ancient Jomon Cedar. If you look up from an opening within the spacious hollow of the stump, you can see the sky cut out in the shape of a heart, which makes it a very popular sightseeing spot.

Ernest Henry Wilson, a British botanist, introduced the existence of this great stump to the world, and hence the great stump is now known as the Wilson Stump. However, documentation on the details of the story behind his discovery, at least in Japanese, is scarce.

Wilson landed in Yakushima one hundred years ago in 1914. What was his life like? What brought him to this isolated island in southern Japan? What were his impressions of the island?

These questions haunted me, and I started studying English references and writing a biography of Wilson for a local magazine two years ago. I found out that the photographs he had taken in Yakushima were preserved at the Archive of the Arnold Arboretum of Harvard University.

In all, there are fifty-seven glass plate negatives of those photographs. They show gigantic trees in the deep forest, granite rocks covered with fresh moss, crystal water rushing down steep slopes, and scenes of quaint villages. They present the island as it was one hundred years ago.

One day as I was immersed in those photos, I noticed some people in the scenery. Looking more closely with a magnifying glass at the tiny figures beside a great tree, I found that they were local young men in kimono standing straight with serious faces. What were their thoughts when they stood in front of the camera that day?

A copy of that photo in hand, I inquired about them to old villagers living near the trailhead. Eventually, I obtained some information, and one by one the names of the three youths in Wilson's photo were revealed. I came to learn that each of them treasured memories of the precious time they shared with Wilson. The entire process abounded in thrills and delights as the mysteries were unveiled one by one.

It cannot be a mere coincidence that this year marks the twentieth anniversary of Yakushima's designation as a World Natural Heritage Site, and next February the centennial of Wilson's visit to the island. It is as if these men have been waiting for this timing to awaken from their long slumber to convey an important message.

I would be overjoyed if this book would serve as an opportunity for readers to consider the significance of the message that Wilson left us in Yakushima.

November 2013
In celebration of the twentieth anniversary of Yakushima as a
World Natural Heritage Site
Tomoko Furui

E.H.Wilson Photographs
@ President and Fellows of Harvard College Arnold Arboretum Archive
写真提供：ハーバード大学

The Sanderson Camera, similar to that which Wilson used in Yakushima.
ウィルソンが屋久島での写真撮影に使用したものと類似の「ザ・サンダーソン・カメラ」

© 日本カメラ博物館 JCII CAMERA MUSEUM

目次 Contents

旅の始まり Beginning of the Expedition ……………………………… 8

屋久島 Island of Yakushima ………………………………………… 14

神の領域へ The Domain of the Gods ……………………………… 20

太古の谷 Ancient Valley ……………………………………………… 26

山の案内人 Forest Guides …………………………………………… 32

サンダーソン・カメラ The Sanderson Camera ……………………… 40

巨木に会う Meet the Old Trees ……………………………………… 46

ウィルソン株 Wilson Stump ………………………………………… 52

命のつながり Endless Succession of Life …………………………… 62

伝説のマツ Legendary Pine ………………………………………… 70

100年後への伝言 Heritage from 100 Years Ago …………………… 82

若者たち、その後 Sequel of the Three Young Men ………………… 84

ウィルソンの歩いた道 Wilson Trail ………………………………… 87

ウィルソンの生涯 The Life of Ernest Henry Wilson ……………… 88

謝辞 Acknowledgments ……………………………………………… 94

参考文献 References ………………………………………………… 95

"If we do not get such records of them in the shape of photographs and specimens,
a hundred years hence many will have disappeared entirely."

Ernest Henry Wilson

もし写真や標本で記録を残さなかったら、
100年後にはその多くは完全に消えてなくなってしまうだろう。

アーネスト・ヘンリー・ウィルソン

Beginning of the Expedition
旅の始まり

"I go to the island of Yakushima where are said to be fine forests of wild Cryptomeria and I am all eager to see them for myself."

野生のスギが生息すると聞いた屋久島に行こうと思う。
そして、私はぜひとも私自身の目でそれを見てみたいと願っている。

「歩き回る日々に本当に疲れた。終わりがくることを願っている。中国各地をただ歩き回るほかは何もしていない気がする」プラント・ハンターになって以来、どんな困難にも果断に立ち向かってきたウィルソンにしては珍しく、弱音ともとれる胸の内を日記に記したまさにその日の午後、突然山が崩れ大きな岩石が彼の足の上に落ちてきた。中国四川省の奥地、4回目の中国探検中での出来事だった。

本の執筆、講演、ラジオ番組への出演そして授与された数々の賞。致命的な傷を負った英雄を待っていたのは、アメリカでの熱狂的な賞賛の嵐だった。

しかし、長くフィールドを生活の場として生きてきたウィルソンの心の中には、どこか隙間風が忍び寄るような満たされぬ思いがくすぶっていた。そんな折、ハーバード大学アーノルド植物園が用意してくれた日本への調査旅行に、ウィルソンは妻と娘を伴っていくことを決心した。家族と共に未知の国へと旅立つのは、初めてのことだった。

"I am certainly getting very tired of the wandering life and long for the end to come. I seem never to have done anything else than wander, wander –through China."

Wilson always faced any difficulty bravely as a plant hunter. He was rarely heard to complain, but on the afternoon of the very day that he entered this in his journal, there was a sudden landslide and huge rocks crushed his leg. The accident occurred in the back country of Sichuan Province on his fourth expedition to China.

When he returned to the US, his time was taken up with writing books, giving lectures, and performing on radio programs. He was awarded numerous prizes. Storms of enthusiastic praise awaited the hero who had sustained serious injury. But Wilson felt uneasy and discontented. He had always lived and worked in the fields. It was at such a time that the Arnold Arboretum at Harvard University offered him the opportunity to go on a research trip to Japan. Wilson decided to take his wife and daughter with him. It was the first time for him to go to an unknown land with his family.

Mrs. Wilson sitting on the rikisha and her daughter Muriel, Karasawa, Kyoto, Japan, 1914.
人力車に乗っているウィルソン夫人とその前に立つ娘のミリエル（左）。京都唐沢にて 1914年

Beginning of the Expedition
旅の始まり

1914年2月3日、予定より2日遅れでウィルソン一家は横浜の港に足を下ろした。情報を得るために訪れた東京帝国大学付属小石川植物園で耳にしたのは、日本の南端に位置する小さな島の名前だった。

屋久島——そこには、太古の巨大スギが今なお野生のまま生息しているという。ウィルソンは急きょ予定を変更して、日本探検の最初の一歩をその島からスタートさせることにした。

妻と娘を東京の帝国ホテルに残したまま、林野管理局の職員、通訳、そしてフィールド・ワーク用の装備一式とともに、勇んで午前8時30分新橋発下関行きの特急列車に飛び乗った。2月14日のことだった。

2年前に開通したばかりの特急列車は、日露戦争に勝利して列強の仲間入りを果たした日本の威信を象徴するものだった。「それは中国の旅とは全く性格の異なる旅で、私にとっては休暇のようなものになるだろう」

洋食を出す食堂車に英語を話す列車長付きという高級感あふれる設備と装飾に包まれた特別仕様の車両は、九州内を走る急行に連結され、東京を出てからわずか36時間ほどで一行を鹿児島まで運んだ。

On February 3, 1914, the Wilsons descended from their ship at the Port of Yokohama two days behind schedule. He visited the Imperial Botanical Garden in Tokyo to obtain information and heard the name of a small island at the southern end of the Japanese archipelago.

The name of the island was Yakushima. He heard that gigantic ancient wild *Cryptomeria japonica* (Japanese cedar known as Yakusugi) still thrive there. Wilson immediately changed his mind and decided to start his expedition in Japan from that island.

Leaving his wife and daughter at the Tokyo Imperial Hotel, Wilson hopped on the limited express to Kagoshima at 8:30 am on February 14. He took along equipment necessary for fieldwork and was accompanied by an officer of the Forestry Bureau and an interpreter.

"This expedition has been totally different in character to those undertaken in China. This trip would be one long holiday for me."

The custom-made posh train, complete with a dining car serving western cuisine and an English speaking train conductor, carried the party to Kagoshima in about thirty six hours.

Wilson(left) and the Japan railway used on the expedition in Japan.
日本探検に使った列車の前のウィルソン（左）。

Beginning of the Expedition
旅の始まり

旅の終着駅に下り立った時、彼の目に映った鹿児島には桜島の大爆発の爪痕が未だ大きく残っていた。

噴煙の高さは1万mを超え、火山灰や軽石、溶岩などの噴出物の総量約20億m³に上ったこの爆発は、噴出量の規模も被害も日本国内では20世紀最大の火山噴火だった。1か月たった後でも火山は活発で余震がまだおさまらず、錦江湾に浮かぶ桜島は興奮を抑えきれない様子で小爆発を繰り返していた。市内の降灰はひどく、いたるところで溶岩の残骸や崩れた石塀が見られた。

なんと、ウィルソンはあの歴史的な「桜島大正大噴火」直後の鹿児島に遭遇していたのだった。

ここで、鹿児島大林区署の優秀な林業技師2名が案内役として合流した。そのなかのひとり、東京農林学校を卒業したばかりの若く情熱的な三好哲男は、ウィルソンにとって屋久島旅行の「最も魅力的で熱心な旅の友」となった。

When he got off the train at the terminal in Kagoshima, he saw terrible scars left by the great eruptions of Sakurajima. The eruptions which began the previous month were the greatest Japan experienced in the twentieth century. Thick smoke rose over ten thousand meters high, and total ejecta including ashes, pumice, and lava amounted to two billion cubic meters. Aftershocks continued even after a month, and the volcano in the bay of Kinko-wan continued to erupt on a small scale. Ash fall and the remains of lava and broken stone walls were everywhere in the city.

Two brilliant forest experts from the Kagoshima Forestry Bureau joined Wilson's team here as guides. One was Tetsuo Miyoshi, an enthusiastic young man, a fresh graduate of Tokyo School of Agriculture and Forestry, and who was to become "a most charming and enthusiastic travelling companion."

View of lava deposits 80 ft. high and devastated Pine forests, Sakurajima, Kagoshima, 1914.
高さ24mの溶岩の堆積物と被害を受けたクロマツ林。鹿児島市桜島　1914年

Island of Yakushima
屋久島

"The southern island of Yakushima is a gigantic upthrust of the igneous rock."

最南端の島、屋久島は巨大な火成岩の隆起だった。

屋久島までは蒸気船「大川丸」が定期運行していた。幸いにも天候に恵まれ、ウィルソンを乗せた船は浮流する軽石をかき分けながら鹿児島港を出て、次の日の朝、2月17日には種子島西之表港に着いた。そこで、荷揚げ作業に時間がかかっている間、一行は上陸して海岸沿いの植物を調査した。プラント・ハンターの触手が動いたのは、マツの木だった。中国奥地の山地で見かけたものとよく似たマツが、平地に自生していたからだ。

昼近く、船はようやく屋久島の玄関口、宮之浦港に向かって蒸気をあげた。皿を伏せたような種子島と対照的に、海から空に向かって聳え立つ屋久島は一目でそのほとんどが険しい花崗岩で構成されている島であることがわかる。

近づくにつれ、頂に雲を抱いた山の全様が徐々に大きくなっていく様子は圧巻だった。鬱蒼とした緑の山並みが海岸まで稜線を延ばす立体的な姿が、目を見張るような美しさで迫ってきた。

A steamship operated regularly to Yakushima. Blessed with fine weather, Wilson boarded the ship which sailed out of the Port of Kagoshima making its way through floating pumice, and arrived at the Port of Nishi-no-omote on the island of Tanegashima the following morning. The party got off the ship and studied plants along the coast while the crew unloaded goods. The instinctive plant hunter in Wilson immediately took note of the pine trees. Wild pines similar to the ones he had seen in the mountains of inland China grew on the flat land here.

It was almost noon when the ship finally departed for the Port of Miyanoura, the gateway to Yakushima. In contrast to Tanegashima which had the appearance of a plate turned upside down, Yakushima soared into the sky, and it was apparent that it was formed by rough granite. He was overwhelmed to see the mountains crowned with clouds growing larger as they approached. Thick green mountains dominated the island stretching out to the seacoast creating a view of astounding beauty.

Village of Miyanoura, 90 miles south of Kagoshima, Yakushima.
鹿児島より144km南の宮之浦集落　屋久島

Island of Yakushima
屋久島

　船は直接陸には着岸できず、手漕ぎのはしけ4、5隻がやってきて人と荷物を港まで運んだ。冬の最中だというのに、たっぷりと水蒸気をふくんだ空気が心地よい暖かさで身体をくるんでくれるようだった。
　屋久島はそのほとんどが山で、村人は山の稜線と海が接するわずかな平地に集落をつくって暮らしている。山のふところに互いに寄り添うように木造家屋が立ち並ぶ小さな村々が、海岸沿いの川の近くに点在し、わずかな耕作地と草原帯がそれを取り囲んでいた。
　丸い石を一面に乗せた板葺の屋根が特徴的だ。その屋根のひとつひとつから暖をとる細い煙が白く立ち上り、木を燃やす匂いが風にのって流れてくる。集落の後ろに見える里山は、人々の生活のための薪、家屋や漁船の資材の調達の場として活用され、また戦争需要の炭焼も盛んだったせいで見事な禿げ山になっていた。長く山に関わってきた人々の暮らしぶりが見てとれた。

　As the ship could not be docked directly to the land, several barges were rowed to the ferry to carry the passengers and loads to the port. Though it was midwinter, the warm atmosphere full of moisture was actually pleasant.
　The island of Yakushima was almost entirely mountains, and the people lived in small villages built on the narrow strips of flat land at the foot of the mountains on the coast. Hamlets of wooden houses clustered together surrounded by limited farmland and grassland lay at the foot of the mountains here and there.
　The houses were roofed with wooden boards topped with round stones. Thin white smoke rose from each roof, and the smell of wood burning for heating drifted in the air. Domestic woodlands, called satoyama, behind the villages were quite bare since the people had taken firewood and building material for houses and fishing boats, as well as for making charcoal for the demands of the Russian-Japanese War. It was apparent that the people relied on the mountains for their living.

Melia azedarach, height 30ft. × girth 7ft. Tree habitat with two men (and small local girl) by road with woodpile. alt.30m

道脇の薪の山とセンダンの木　樹高 9m×周囲 2.1m　薪が積まれた道のそばに 2 人の男とその横に少女の姿が見える。標高 30m

Island of Yakushima
屋久島

　宮之浦川沿いの旅館で一行にあてがわれた部屋は、総屋久スギ造りの二間続きの和室だった。千年以上の年月を経た木だけがもつ独特の芳香と、赤みを帯びた木目から放たれる光沢がウィルソンの旅の疲れを癒してくれたことだろう。

　川に面した縁側からは、対岸に水田が見えた。島にはまだ電気はない。夜の闇が訪れると、白い霧が流れ始め、家々からかすかに漏れるランプの灯火が、黒い川面にわずかな光を投げかけていた。

　西洋人の客を迎えるのは、この宿にとっては創業以来初めての歴史的な出来事だった。ウィルソンが泊まった部屋は、『ウィルソンの間』と名付けられ、近年まで大切に保存されていた。残念ながら、100周年を目の前にして維持管理の困難から取り壊されてしまったが、地元の有志により柱、壁、床、天井、欄間などの部材がすべて引き取られ、現在別の場所での再建が試みられている。

At an inn along the Miyanoura River, the party stayed in a guestroom comprised of two tatami rooms which were divided by sliding partitions. The guestroom was made of Yakusugi aged over a thousand years. The distinct aroma and polished fine red grain of this wood must have been comforting to Wilson.

From the wooden passageway of the inn, they could see rice fields after harvest on the other side of the river. Electric power had yet to reach the island. With the dusk, white mist began to emanate, and weak lamp lights leaking from the houses cast dim reflections on the dark river surface.

The inn had never accommodated a westerner before, and Wilson's stay became the most historical event since its establishment. The guestroom Wilson stayed in was named "Wilson-no-ma" and was preserved until recently. Due to difficulties in maintenance, the inn was torn down before the centennial of his stay. However, some local people volunteered to keep all the furnishings including the pillars, walls, floor, ceiling and railings to rebuild the guestroom elsewhere.

View of cone-shaped bald hills with village of Shitogo and rice-fields.
志戸子集落と田園風景。背後に円錐形の禿げ山が見える。

The Domain of the Gods
神の領域へ

There are no open spaces, but everywhere dense primeval forest.

開けた空間はなく、どこもかしこも密な太古の森が広がっていた。

翌朝、日が昇るとともに屋久島小林区署の職員が宿にやってきて、山に登る段取りが整えられた。総勢6名となったキャラバン隊は山に登るべく宮之浦の里を後にした。

山が険しく雨量の多い屋久島では、200本を超える大小の川が傾斜を駆け下り海に流れ込んでいる。この頃大きな河川にはまだ橋がなく、集落間の移動は主に小さな舟が使われていた。登山口のある隣り村に行くため、一行は宮之浦河口近くを繰り舟で渡った。

水田に続く草深い草原を通り抜けると、楠川と呼ばれる牧歌的な佇まいを見せる村に出た。家々を囲む石垣と掘割を流れる清い水の流れが小奇麗な風景を創り出している。水汲み、野菜洗い、魚さばきなど家の前の川で朝の仕事に勤しんでいた村人たちは、青い目の外国人の出現に腰を抜かさんばかりに驚いた。子供たちは一様に口を開け目を丸くして化石のように立ち尽くし、通り過ぎる一行をいつまでも見つめていた。

At sunrise the following morning, a Yakushima Forestry officer came to make arrangements for climbing the mountains. The caravan of six members left the village of Miyanoura for the mountains.

As Yakushima is all steep mountains with much rainfall, it has more than two hundred rivers and streams running down its slopes into the sea. In those days, there were no bridges over the big rivers, and small boats were mainly used to go to other villages. The party took a boat to cross the Miyanoura River near its mouth to go to the neighboring village at the trailhead.

They passed through rice paddies and deep grassland and reached the picturesque village of Kusugawa. It was a neat village with houses surrounded by stone hedges and clear water flowing down small canals. Villagers were outside tending to their morning chores drawing water, washing vegetables, and preparing fish in the canals in front of their houses. They were utterly astounded to see a foreigner with blue eyes leading officers in uniform. Children stood paralyzed with their mouths and eyes wide open and stared as the team passed by.

Ficus Wightiana, height 50ft. × girth 10ft. Sea-level.
海岸近くのアコウの木　樹高 15m × 周囲 3m

The Domain of the Gods
神の領域へ

　ガジュツ、サフランなどの薬草の栽培地が続く緩やかな坂の先の山道は、屋久島で最初につくられた古い登山道で、当時は深山に分け入る唯一の道だった。また古くから岳参りのルートとしても知られ、山頂の祠に詣でるための信仰の道でもあった。ヘゴ、シダなどが繁茂するなだらかな草原に続き、植林されたスギが林立している森林地帯に入ると、ところどころに炭焼の窯が見えた。

　この地点に「野のヨケイ」と呼ばれる小さな祠がある。これは人の領域と神の領域を分ける聖域境界の印のようなものである。屋久島では奥山は神が住むところとされ、江戸時代中期まで、村人は決して奥山の老樹に手を触れることはなかったのだ。

　ここで聖域へと向かう巡礼者に新しい仲間が加わった。山の案内、荷物の運搬、食事の用意などのために雇われた地元の若者数名だった。若者たちは揃って綿入れの着物に脚絆、地下足袋という装いて現れた。常から営林署の仕事を、度々手伝っていたのだろう。祠に手を合わせた後に、手慣れた様子で背負子に荷物を乗せると神の領域へと足を踏み出した。

　現存するウィルソン手書きのフィールド・ノートは、この時点2月18日の朝から始まっている。

　The Kusugawa mountain trail was the oldest trail on Yakushima, and the only way to go into the deep mountains. It long served as a religious road for those who went up to worship the gods of the mountains enshrined at a small shrine at the peak. First they went through fields covered with tree ferns and ferns. Then appeared a forest of artificially planted cedars with kilns for baking charcoal here and there.

　A small shrine stands there. It marks the boundary that separates the domain of humans from that of the gods. In Yakushima the deep mountains were considered as the domain of the gods, and the islanders never touched the ancient trees in the deep mountains until the middle of the Edo era.

　Several local youths joined the team of pilgrims heading for the sanctuary. They were hired to guide them up the mountains, carry their loads, prepare meals and tend to other tasks. These youths were all clad in quilted kimonos and cotton trousers and wore footgear made of cloth. They seemed to be used to working for the forestry office. After praying at the shrine, they quickly loaded their back carriers and stepped into the domain of the gods.

　Wilson's handwritten field notebook, which still remains today, starts at this point on February 18, 1914.

Gleichenia glauca. Vegetation on hillside. Tree fern at extreme left. Savannah region.
里の草原地帯。傾斜地のウラジロ。左端にヒカゲヘゴが見える。

The Domain of the Gods
神の領域へ

平地が谷間に変化し、川沿いの歩道は峡谷に向かって狭まりながら上り始め、同時に次第に森は深く密になっていった。やがて、急峻なジグザグの登りになるにつれ、樹間から流れる風が急に冷たくなり、風景は針葉樹が立ち並ぶ森林に姿を変えた。奉行歩道とも呼ばれるこの山道は、藩政時代は年貢の代わりの杉の平木や板木を運ぶ村人の生活の糧の道であった。

ひとつひとつ丁寧に積み上げ、定期的に整備もされてきたのだろう、七重八重の曲がり坂には滑らないように少し上向きに配置された石の階段が続いていた。途中には屋久スギの分厚い板で作られた水槽と竹の柄杓が設置された、巡礼者のための水飲み場もいくつかあった。

標高が上がるにつれて石段は苔の衣をまとい始め、森の大気さえ緑の濃淡のベールのように染まって見えた。傾斜を彷徨うように登った先、川音が小さくなったあたりに三本杉の巨大な山門があった。神の森への入り口だった。

The plains gradually turned into valleys, and the riverside path began to narrow as it headed up into the forest which grew dense. The path became steeper as it zigzagged up the mountain, the winds blowing through the trees suddenly got cold, and they were now in a forest of conifers. In the feudal period, villagers carried cedar shingles and lumber along this trail to earn their living and pay their taxes in the form of shingles. Thus, the trail was also called the path of the magistrates.

Old stone steps which had been carefully laid and regularly maintained, and tilted upwards to prevent slipping, continued on the winding slope. Occasionally, water tanks made of thick Yaku-sugi boards with bamboo dippers provided drinking water to the pilgrims along the road.

As the altitude rose, the stone stairs became covered with moss, and even the forest atmosphere seemed to be dyed in a veil of various shades of green. As they wandered up the path, and the river could hardly be heard, they were met by three towering great cedars which served as the entrance to the forest of the gods.

Stand of trees with path. *Quercus myrsinaefolia*, height 45ft. × girth 5ft. *Cryptomeria japonica* (Japanese cedar) behind.
林立する木と小道。シラカシ　樹高 13.7m ×周囲 1.5m　後ろにスギ。

Ancient Valley
太古の谷

The wood of fallen Trees is made into shingles and carried out on men's backs, but that is all that is being done at present.

倒れた木は屋根材にされ、人間に背負われて運び出される。それが現在、行われているすべてだ。

　針葉樹の森の中を縫うように続く神の道は、何千年もの間の滴が育てた緑苔に厚く覆われた静かな参道だった。道に沿って流れていた川はいつしか細い渓流となり、苔むした岩棚を滑り落ちていた。丸太の橋の下を流れる狭い川床を横切ると、モミ、ツガ、ヒメシャラなどの大木が目に入るようになってきた。

　島の若者たちはカメラの装備や、食料品などを背負い、着実な足取りで黙々と歩みを進めていく。そのなかのひとり、ひときわ背が高く肩幅が広く逞しい体つきをした若者が、調理用具のようにも見える鉄製の釜を背負っていた。中ほどが丸く膨らんでいるので、ほとんど体を海老のように曲げて歩く様子にウィルソンは目を見張った。

　13歳から大人に混じって山仕事を始め、重い荷を運んで山の斜面を上り下りして過ごしてきた彼らは、まさに働き盛りの山男だった。謙虚でひかえめで、寡黙に仕事をこなすその姿に、ウィルソンはいつしか若い頃の自分を重ね合わせていた。

The path of the gods that wove through the grove of conifers was a serene path covered with thick verdure mosses nurtured by dew over thousands of years. The river flowing along the path was now a mountain stream sliding down mossy boulders. When they crossed the narrow riverbed over a log bridge, tall Japanese fir, Japanese hemlock, and Japanese stewartia came into view among the mossy forest.

The island youths with camera equipment and food supplies on their backs proceeded quietly in an accustomed fashion. On the back of one lad was an iron cauldron so huge that Wilson wondered if it was to be used for cooking. The lad was particularly tall, had broad shoulders and a strong build. Wilson's attention was caught by the sight of him hunching like a lobster as the cauldron bulged in the middle.

These young men had begun working in the mountains among the grownups, starting from the age of thirteen. Carrying heavy loads up and down the mountain slopes, they were now mountain workers in their prime. Modest and humble, they attended to their tasks in silence. They reminded Wilson of his own youth.

Two men standing on bridge of single log. At right Tetsuo Miyoshi(right), Kagoshima forestry officer, with *Cryptomeria japonica* (Japanese cedar) forest in the distance.
丸太橋。向かって右側の男性が三好哲男(鹿児島大林区署職員)。遠くにスギ林が見える。

Ancient Valley
太古の谷

標高970mの峠に差し掛かり周りを眺めると、深い雪に覆われた白い峰々が見えた。南国の島とはいえ、屋久島の山の頂はすっぽりと2月の装いに包まれていたのだ。ウィルソンの負傷した足は未だ完全ではないことから、山の深部にまで行くことは断念しなければならなかった。

峠を越えて南斜面に出ると打って変わって苔は少なくなり、爽やかな二次林からなる森に入った。やがて、安房川の川岸と合流する通称「楠川分かれ」というT字路にぶつかると、森のなかにしては珍しく開けた場所に出た。標高600mから1500m、東西7km、南北6kmにわたる大きな盆地状の地形をしたそこは、地図上では屋久島のほぼ中央にあたる。

江戸時代には、薩摩藩公用の材木が伐採され保護管理されていた場所で、「小杉谷」とも呼ばれていた。現在、小杉谷集落跡があるところよりはさらに30分ほど奥の地点にあたる。

そこは、ウィルソンが一目見たいと切望した樹齢数千年を超える太古の樹林、屋久スギが多数生息する場所であった。

White peaks covered with deep snow came into view as they approached a ridge at an altitude of nine hundred seventy meters. Though Yakushima is in the south, its mountaintop was robed for February. As his injured leg had not fully recovered, Wilson had to give up going into the heart of the mountains.

The scenery on the southern slope after crossing the ridge was quite different with less moss and a breezy secondary forest. Gradually they came to a T-junction where the path hit the Anbo river bank, and reached a wide clearing, a rarity in forests. In the form of a large basin, its altitude ranged from six hundred to one thousand five hundred meters, and stretched seven kilometers wide east to west and six kilometers long north to south almost in the center of Yakushima.

In the Edo period, lumber was logged for the feudal lord of Satsuma Province, and the site was protected and managed by the province. This was the ancient valley that Wilson had wished to see with numerous Yakusugi which had thrived for several thousand years.

Abies firma (Japanese fir), height 80ft. × girth 12ft. in the foreground, *Trochodendron aralioides*, height 50ft. × girth 6ft. alt.780m

モミ　樹高 24m ×周囲 3.7m　前にヤマグルマ　樹高 15m ×周囲 1.8m。標高 780m

Ancient Valley
太古の谷

盆地の川沿いに、営林署の署員が使う小屋があった。ウィルソンはそこをベースキャンプにして、周辺の森を探索することにした。

1879年（明治12年）の新政府による地租改正によって、屋久島の山林はほとんど国有林となった。島民にとって山は誰のものでもない、神のものだった。その神から、生きるために必要な物資をいただいて暮らしてきた。それが急に山に入って木を取ってきたものは、盗伐の罪を問われることになったのだ。合点がいかない島民が立ち上がり、裁判を起こしたのも頷ける。

ウィルソンが屋久島に来た1914年の時点では、まだ裁判が結審されていない。この頃、山の事業は営林署によって細々と行われ、屋久スギの倒木、切株などが公売されていた。小屋の周りには、玉切りにされた木、製材した板木などが散らばっていて、手斧を使った人力による伐採の様子がうかがえた。

Along the river in the valley stood a hut used by forestry officers. Wilson decided to set his base camp here for exploring the surrounding forests.

With the Meiji Restoration, the new government had revised the land tax system in 1879, and designated most of the mountain forests on Yakushima as national property. For the islanders, the mountains did not belong to anyone; they belonged to the gods. The gods provided whatever items were necessary for the people to live on, but under the revision, whoever took trees from the mountains would be considered a thief. It is quite understandable that the islanders could not consent to this and took the case to court.

In 1914, when Wilson came to Yakushima, the trial was not yet concluded. Forestry was operated on a very small scale by the forestry office. Fallen trunks and stumps were sold at public auction. Cross-cut logs and lumber were scattered around the hut, showing that hand tools were used for logging.

Cryptomeria japonica (Japanese cedar), height 80ft. × girth17ft. with a shed and randomly piled lumber in the foreground. alt.780m

スギ　樹高 24m ×周囲 5.2m　前に作業小屋と周囲に散らばっている板木。標高 780m

Forest Guides
山の案内人

There are no roads, only trails laid on logs for the use of the forests guards, and no accommodation other than rough huts for the occasional use of these men.

森林官が使用するための丸太を敷いた手づくりの道があるだけで、登山道はなかった。
また彼らが使う簡素な小屋よりほかに宿泊施設もなかった。

山での最初の一日が終わろうとしていた。盆地を囲む濃い森は沈黙の中に身を沈め、柔らかな影を足元に落とし始めていた。

島の若者のなかでリーダー格と思われる3人が、背負ってきた釜を即席のかまどに据え置くと川水を汲みいれ、薪で釜の底から直火で温め始めた。やがて頃合いを見て、ウィルソンにその中に入るように促した。白い湯気がもうもうと立つ釜を真ん中に、逞しい山男たちが無愛想に立っているのだ。

「おいおい、私を煮殺して食べるつもりなのかい？」

ウィルソンが声をあげると、それを通訳から聞いた若者たちははじけるように豪快に笑い出した。

若者たちがウィルソンのために用意したのは、実は五右衛門風呂だったのだ。野外の冬の夜にはまったく贅沢なもてなしだった。以後ウィルソンの必需品となり、毎夕森林を眺めながら、ぬるめの湯に長時間入るのを好んだという。

遠来の客のために、鉄の釜を背負って山道を登った若者の名は、渡辺比賀之助といった。

The first day in the mountains was almost finished. The deep forests surrounding the valley stood in solemn silence, and started to cast soft shadows.

Three young men of the island, who appeared to be leaders, set up the large cauldron on a makeshift fireplace, poured river water into the cauldron, and started to heat its bottom with firewood. After a while, they asked Wilson to get into the cauldron. The well-built mountaineers bluntly stood around the steaming cauldron.

When Wilson asked half seriously through the interpreter,

"Wait a minute. Are you going to boil me to death and eat me?", the young men burst into loud laughter.

The cauldron was popularly known as "goemon-buro", a bathtub commonly used in households on the island. The bathtub represented great hospitality for a guest who had to spend nights outdoors in winter. The bathtub turned out to be a must for Wilson, who loved to take a long warm bath while enjoying the sight of the forests every evening.

Higanosuke Watanabe was the young man who carried the iron cauldron on his back climbing along the mountain path for the visitor from afar.

Distylium racemosum (leaning), height 60ft. ×
girth 10ft. *Trochodendron aralioides*, height 55ft. ×
girth 4ft. with an open shed in the foreground.
alt. 780m

傾いているイスノキ　樹高 18m ×周囲 3m、
ヤマグルマ　樹高 17m ×周囲 1.2m　前にむ
き出しの小屋。標高 780m

Forest Guides
山の案内人

　比賀之助より2歳ほど年上で明るく愛嬌があり、人懐っこい若者がいた。名を牧次郎助といった。まだ幼さの残る童顔で、そのせいか十代の少年のようにも見える。次郎助はウィルソンのどんな動きも見落とさないように観察し、意味がわからなくとも言葉を聞き洩らさないように耳を澄ましていた。そのせいか、ウィルソンはいつも彼をそばに置いていたという。
　次郎助の主な役目は食事の準備だった。ウィルソンが特に気にいったのは、森から採ってきた山菜の天ぷらと川で仕留めたウナギのかば焼きだった。
　「この島には自然のなかに食べられるものが豊富にある。まったく素晴らしいところだ」
　食後に、次郎助は自宅で母親がつくった手もみの緑茶を差し出した。すると、ウィルソンはバッグから小さな四角い白い塊を取り出して茶のなかに入れた。白砂糖だった。サトウキビを煮てつくる黒砂糖しか知らなかった次郎助には、その砂糖の白さと小枝で茶碗のなかをかき回すウィルソンの奇妙な仕草が、生涯忘れられない思い出となって心に残った。

　There was another young man about the same age as Higanosuke. His name was Jirosuke Maki. He was a cheerful, charming, and friendly man. Due to his baby face, he looked like a teenager. Jirosuke carefully observed every movement that Wilson made; he listened intently and tried not to miss a single word, though he did not understand what Wilson was saying. Wilson allowed him to closely accompany him.
　Jirosuke was mainly responsible for preparing meals. Wilson was particularly fond of tempura of wild vegetables picked in the forests and grilled eels caught in the river. Wilson thought, "The nature of this island offers abundant food. This is a truly amazing place."
　After a meal, Jirosuke served green tea made from home-made tea leaves prepared by his mother. Wilson took out a small cube from his bag and put it into the tea. It was a lump of white sugar. Jirosuke knew only of brown sugar made by boiling down sugarcane juice. He was impressed, for the rest of his life, by the whiteness of the sugar and Wilson's strange action of stirring the cup with a twig.

Cryptomeria japonica (Japanese cedar) with a spiraling trunk, height 80ft. × girth 17ft. A pile of lumber and frame made of slender tree trunks in the foreground. alt.780m

ねじれた幹をもつスギ　樹高24m×周囲5.2m　前に製材された板と細木でつくった柵が見える。標高780m

Forest Guides
山の案内人

リーダー格の3人の若者のなかで、いちばんのはにかみ屋は大石喜平だった。痩せた細身の体で、時として寂しい印象すら与える若者だった。比賀之助のようなふてぶてしい逞しさも、次郎助のような快活な性格も持ち合わせていなかったが、喜平は黙って真面目に仕事をこなすタイプだった。骨惜しみしないその姿は気立ての良さを感じさせた。

喜平はまた、詩人だった。ちょっとした風景の変化、風の動き、石の割れ目に咲く小さな高山植物の色合い、そういったものに心を奪われている様子が、度々見受けられた。

山のなかで興味を引かれる木を見かけると、ウィルソンはよく幹に耳を当ててじっと目をつぶっていた。それをためらいがちな目で見ていた喜平は誰も見ていないときに、こっそり真似をして同じように幹に耳を当ててみた。すると、幹の内部を流れる水音がした。

生き物の息遣いが鼓膜を通して心に響いた。この新しい発見に、喜平は小躍りしたいような喜びを感じたという。

Of the three young men, Kihei Oishi was the most bashful person. This slender young man sometimes looked lonesome. He was not as tough or muscular as Higanosuke, nor was he as cheerful as Jirosuke. He was the kind of person who worked diligently and quietly. His attitude represented devotion and modesty.

Kihei was a poet, too. He was often seen fascinated by small changes in the landscape and wind, as well as the color tones of tiny flowers of alpine plants growing in rock crevices.

When Wilson spotted trees of interest in the mountains, he often put his ear on their trunks with his eyes closed. Kihei, who liked to observe what Wilson was doing, put his own ear on a tree trunk when nobody was around; he heard the sound of water flowing in the trunk.

The workings of living things touched his emotions through his ear. New discoveries gave Kihei a sense of delight and joy.

Cryptomeria japonica (Japanese cedar), height 65ft. × girth 15ft. Old tree, top broken off. alt.780m

老齢スギ　樹高 20m ×周囲 4.6m　先端部が折れている。標高 780m

Forest Guides
山の案内人

ウィルソンが1日2円の賃金で雇った若者たちのなかで、楠川村の比賀之助、次郎助、喜平の3名が全行程について歩き、時には写真のモデルになりながら植物採取の手伝いや日常身辺の世話をしたと思われる。あとの若者は交替で里に下りて食料などの補給にあたったようだ。楠川村に伝承が残っていないことから、他の村からの応援要員だったのかもしれない。

3名の若者はウィルソンと過ごした日々を子供たちに語り、子供たちはまたその子供たちに伝え、まるで家族の伝説のように一連の物語が今も里に残されていた。

一行は寝る前に焚火を囲んで座り、地元の手造りの焼酎を酌み交わした。ウィルソンは島の若者たちにヨーロッパやアメリカや中国など遠い国の話を語った。そして、夜がふけ焚火の炎が小さくなりかけた頃に、必ずこうつけ加えたという。

「しかし、屋久島の山ほど豊かな素晴らしい山はほかにはなかった。だから、君たちのような若い人たちがこの山を守っていってほしい」

この言葉は、山の案内人たちの魂に深く染み入った。そして、それぞれがウィルソンの姿を心のどこかに宿しながら、その後の人生を歩むこととなった。

Among the young men who Wilson hired at the wage of two yen per day, three of them, Higanosuke, Jirosuke, and Kihei from Kusugawa, walked with Wilson throughout the journey. Sometimes serving as photo models, they helped Wilson pick plants and took care of him in daily life.

The young men later talked to their children about Wilson. These episodes have been handed down from one generation to the next. A series of episodes are still alive in the village, just like family legends.

Before going to sleep, the party formed a circle around the campfire, and drank shochu, distilled sweet potato liquor made in the village. Wilson talked to the young men about faraway countries including Europe, the U.S., and China. When the fire grew smaller late at night, Wilson always added:

"But no other mountains are as rich in nature and splendid as those on Yakushima. I hope that young people like you will protect the mountains."

These words deeply impressed the mountain guides, who were influenced by Wilson in many ways for the rest of their lives.

Tsuga Sieboldii (Japanese hemlock) leaning over a rocky streambed of Anbo river.
Cryptomeria japonica (Japanese cedar) to left. alt.700m
岩の多い安房川の流れに傾いて立つツガ。左にスギの森が見える。標高700m

The Sanderson Camera
サンダーソン・カメラ

There is no record of anyone –foreigner or native- having previously enjoyed such opportunity to study these plants.

外国人にしろ、日本人にしろ、過去に誰もこれらの植物を調べる機会を楽しんだ人間の記録はない。

ウィルソンが屋久島の山に持ち込んだ荷物のなかで、最もかさばったものはカメラの装備一式だった。彼は、イギリスのサンダーソン社が開発した最新鋭の箱型蛇腹式のフィールド・カメラを使用した。このカメラは、建築物を撮影するために考案されたもので、上面のフラップが開きレンズが上がる仕組みになっている。野外で背の高い木を写すのには最適であった。

しかし、問題はその重さと使い勝手の悪さだった。6.5 インチ× 8.5 インチのガラス乾板多数とそのホルダーのための大きな箱が数個、さらには頑丈な木製の三脚も同時に運ばなくてはならなかった。屋久島で彼がなぜ多くの若者の手を必要としたかが、これで納得できる。

さらに撮影の手順も生半可のものではなかった。ガラス乾板は光が当たると感光するため、ケースの開封やホルダーへのセットなどのすべての作業をバッグの中で手探りで行わなくてはならなかったからだ。

Wilson used a box-shaped bellows-type field camera of the latest model being made by Sanderson in England. Designed to photograph tall buildings, the camera had a mechanism to open the upper flap and lift the lens. Indeed, the camera was suited to photograph tall trees outdoors.

A set of camera equipment was the bulkiest piece of luggage that Wilson brought into the mountains. Beside the camera there were several light proof boxes holding glass photographic plates measuring 6.5 × 8.5 inches, and more boxes of holders for each plate as well as a heavy-duty wooden tripod. This explains why Wilson needed the help of many young men.

The photographing process was also laborious. Glass photographic plates were useless if they were exposed to light. Thus, handling the glass plate into its holder had to be done inside a bag.

Stump and fallen trunk of *Cryptomeria japonica* (Japanese cedar) with young trees, height 40-50ft. × girth 5-7ft. growing thereon. alt.950m
屋久スギの切株と倒れた幹。樹高 12 〜 15m ×周囲 1.5 〜 2.1m ほどの大きさの若いスギがその上に育っている。標高 950m

The Sanderson Camera
サンダーソン・カメラ

　長時間の露光が必要とされるガラス乾板のカメラは、スクリーンやカメラ本体がずれないように細心の注意が必要とされた。シャッタースピードは5〜8分だったというから、完全に無風状態の瞬間を捉えなくてはならない。また、対象物の大きさを示すためにモデルとして樹木の横に立たされた人たちも、彫像のように不動の姿勢をとり続けなくてはいけなかったわけだ。

　この儀式の間に、風が枝や葉を揺すって木を動かしてしまったり、太陽の位置が移動して光が変化してしまうこともあっただろう。また、天候が著しく変わる屋久島の山の中では、予期せぬ雨や霧に見舞われることもあったはずだ。にもかかわらず、10日間の滞在で57枚の写真を撮影したウィルソンの忍耐強さとプロ意識には驚かされる。

　今も鮮明な画像で残されているウィルソンの写真の一枚一枚は、大きな努力の賜物であったのだ。

Cameras using glass photographic plates required long exposure time. Maximum care had to be taken to eliminate any movement of the camera. Because the shutter speed was five to eight minutes, completely windless moments had to be captured. People who were asked to stand next to trees as models to show the size of subjects had to stand motionless like statues.

During such "ceremonies," the wind must have shaken branches, leaves, and even trees. The light must have changed due to the movement of the sun. The mountains of Yakushima, where the weather is highly changeable, must have brought unexpected showers. The 57 photographs taken during his ten-day stay demonstrate Wilson's remarkable patience and professional commitment.

The photographs, which are still vividly retained today, represent Wilson's great efforts.

Cryptomeria japonica (Japanese cedar), height 90ft. × girth 25ft. Note the size of tree as contrasted with a man standing by the trunk. alt 700m

巨大なスギ　樹高 27m × 周囲 7.6m　木の根元に立つ人物との大きさの対比に注目。標高 700m

The Sanderson Camera
サンダーソン・カメラ

写真撮影の現場においては、ウィルソンはあきれるほどの完璧主義者であった。撮影したい木や風景に出会うと、その周りを異なる視点から見てまわり、さらに幹、枝の構成そして背景など最終的に満足が得られるまで何度も確認してからカメラをセットした。

ガラス乾板は壊れやすく取扱いが大変だが、大量のデータを高品質に記録できるという利点がある。ウィルソンが撮った写真を拡大して眺めていると、そこにさまざまなものが同時に映し出されているのに驚かされる。目の前にある「真実」をただ記録するだけではなく、それがそこに存在している「状況」をも同時に捉えようとした撮影者の思いが伝わってくるようだ。自然に対する深い尊敬と愛だけではなく、モデルとなった人物の服装や佇まいにも同時に心を動かされている様子が感じられる。

ウィルソンは撮影が終わったガラス乾板を慎重に荷造りして東京からロンドンに送り、後日すべてを自らの慎重な管理の下で現像したという。

In photography, Wilson was an incredible perfectionist. When he encountered trees or landscapes that he wished to photograph, he walked around to see them from different angles. He checked the arrangement of trunks and branches as well as the background many times over until he was fully satisfied before setting up the camera.

Though fragile and difficult to handle, glass plates offer the advantage of recording a large volume of data with high resolution. One is surprised to find the various things that can be identified on the magnified photographs. The works reflect the photographer's intention to record not just the facts but also to capture the entire situation; the photographs embody Wilson's deep respect and love of nature as well as his consideration for the young men of the island who served as models.

After taking the photographs, Wilson carefully packed the glass plates, shipped them from Tokyo to London, to await his supervision of their development.

Man standing on old stump beside large trees. *Cryptomeria japonica* (Japanese cedar), height 75ft. × girth12 ft. alt.780m

樹高 23m×周囲 3.7m の大きなスギの横、古い切株の上に男が立っている。標高 780m

Meet the Old Trees
巨木に会う

The high humidity and abundant precipitation are very favorable to the growth of coniferous plants. The trees I saw on Yakushima impressed me most.

高い湿度と豊富な降水量は、針葉樹の成長に非常に適している。屋久島で見た木々が最も印象的だった。

屋久島の温暖な気候、豊富な雨、山の険しい傾斜といったその自然の条件すべてが、樹木が育つ最適な環境をつくっていた。
「ほんの小さな島だが、日本で最も興味深く、驚かされる森は、この屋久島の森だった」
ウィルソンは、標高800mから1100mの森を探索しながら、その豊かな植生に驚嘆の声を上げ、興奮した様子でアメリカの友人への手紙に記している。
森の深部にはさまざまな巨木がそびえ立ち、何千年、何百年と生きてきたものだけがもつ威厳を保ちながら、その存在を誇示していた。
モミ、ツガ、ヒノキといった針葉樹が、ヤマグルマ、ヒメシャラなどの広葉樹と仲良く共生する混合林のなかで王者のように君臨するスギは、明らかに人の手で植えられたものではなく、自然が時間をかけて生命を引き継いできたものに間違いなかった。

All the natural conditions of Yakushima, including the warm climate, abundant rainfall, and steep mountain slopes, created an ideal environment for trees to grow.

On his expedition, Wilson was amazed by the rich vegetation of forests located eight hundred meters to eleven hundred meters above sea level. He wrote to one of his friends in the U.S. with excitement.

"It is only a small island, yet the most interesting and remarkable forest in all Japan is this on Yakushima."

Japanese cedar reigned like a king in an environment where conifers such as Japanese fir, Japanese hemlock, and Hinoki cypress lived in harmony with a variety of broad-leaved trees. Obviously, these trees were not planted by people; they were nurtured by nature.

Deep inside the forests stood millennia or centuries-old huge trees of various species; these trees had a unique presence and dignity.

Tsuga Sieboldii (Japanese hemlock), height 75ft. × girth 18ft. with *Hydrangea petiolaris*. *Abies firma* (Japanese fir) behind. alt.780m
ツルアジサイが絡んだツガ　樹高 23m×周囲 5.5m　後ろにモミが見える。標高 780m

Meet the Old Trees
巨木に会う

「屋久島ではツガの素晴らしい生息が見られ、その姿は絵のように美しい」とウィルソンが最大限の賛美を贈ったツガは、高さは26m、幹の周囲が6mほどもある大木がベースキャンプ周辺にはたくさん生息していた。どの木も何かしら傾いていて、船のマストのようにまっすぐに立つスギとのコントラストは目を見張るものがあった。

モミもまた、標高500mの険しい傾斜地から山頂近くの深山にまで生息していて、印象的な姿を際立たせていた。

そして、緑の衣をまとった針葉樹のなかでひときわ目立ってウィルソンの目を捉えたのは、橙色の明るい滑らかな木肌を見せるヒメシャラだった。まるでゴシック建築の楼門のごとく枝を左右に優雅に広げ、霧のなかにすらりと佇むその美しい立ち姿は、森の貴婦人のような気高さと気品にあふれていた。

"On Yakushima, where the finest development of Tsuga sieboldii(Japanese hemlock) is fond, the trees are picturesque in Appearance."

As described by Wilson with utmost admiration, large trees measuring about 85ft. in height and 20ft. in circumference grew in large numbers around the base camp. All of these trees were tilted to some extent, in a sharp contrast with Japanese cedar that grew upright like ships' masts.

The habitat of Japanese fir ranged from steep slopes five hundred meters above sea level to near mountaintops. These trees stood in an imposing manner.

Among conifers clad in green leaves the broad leaved Japanese stewartia drew Wilson's attention with its striking presence. Characterized by smooth, orange bark, the trees elegantly stretched out their branches, just like tower gates in Gothic architecture. The slim trees rising high into the foggy sky were characterized by refinement and gracefulness. In a figurative sense, they were "noblewomen" in the forests.

Stewartia monadelpha bark smooth, pale brown.
Trunk girth 10.5ft. alt.780m
ヒメシャラ　樹皮は滑らかで淡い茶色をしている。周囲 3.2m。標高 780m

Meet the Old Trees
巨木に会う

屋久島は、スギの南限地だった。そのなかで、千年以上の樹齢をもつ老樹だけが屋久スギと呼ばれている。幹には狭い割れ目が入っていて、堅い。幹の皮は赤茶色で、長い年月風雨にさらされた箇所は灰色になっている。まさに驚異的ともいえるその風格は、独特の雰囲気をかもし出していた。

ウィルソンが見た屋久スギの最大樹高は約40mで、平均的な木でも30〜35mの高さがあった。老齢スギのほとんどは頭部を失っていて、幾度となく強風に打ち勝ってきた年月の長さが感じられた。上部の枝は水平に広がり、美しい楕円形の樹冠を形作っていた。

樹脂分を多く含む屋久スギは簡単には腐らない。何百年も前に地面に横たわった倒木や、切り倒された後の切株でさえ、苔むした木肌の下には若木のような色合いを失うことはなかった。

ウィルソンがこの森を見た9年後のことだった。安房川沿いに森林軌道が敷かれ斫伐事務所が開設されて、屋久スギの本格的な大量伐採が始まった。ウィルソンが写真に撮った森の老樹たちの姿は、今はもうほとんど見ることはできない。

Yakushima is the southernmost habitat of Japanese cedar. Trees more than one thousand years old are called Yakusugi. The thick bark of Yakusugi has narrow cracks. The color of the trunk is reddish brown, and areas exposed to weather for many years are gray. Their phenomenal presence produced a unique atmosphere.

The tallest specimen that Wilson saw measured about 131ft. in height. Even ordinary trees are 100-115ft. tall. Most old trees had lost their top parts, which testified to the long history of these trees to surviving strong winds over and over again. The branches in the upper part of the trees stretched out horizontally, creating beautifully shaped oval crowns.

Rich in resin content, Yakusugi does not easily rot. Trees that fell hundreds of years ago and stumps of felled trees retained the color of young trees beneath the mossy surface.

Nine years after Wilson saw the forests, a forest railway was constructed along the Anbo River, and a logging office was opened to launch full-scale logging of Yakusugi. The old trees in the forests photographed by Wilson are mostly gone.

Biggest *Cryptomeria japonica* (Japanese cedar) which Wilson saw with Mr.Yoshitomi, Yakushima forestry officer, who discovered this tree, girth 30ft. alt.1030m
ウィルソンが見た最大の屋久スギ　幹の周囲9m。前に立つのは発見者の吉富友一（屋久島小林区署森林主事）。標高1030m

Wilson Stump
ウィルソン株

*Nowhere else, not even on famed Mt. Omei, in western China,
have I seen such a wealth of his vegetation.*

とこにも、西中国の有名な峨眉山でさえ、私はこのような豊かな植生に出会ったことはない、

2月19日と20日に合計18枚の写真を撮影した後、翌21日にはウィルソンは1枚の写真も残していない。雨が降っていたのか、風が強かったのか。あるいはこの時期、この辺りでは雪が降り積もることも珍しくはないことから、雪交じりの荒れた天候であった可能性もある。ウィルソンは撮影装備を持ち歩くことは断念し、その日は辺りの植物採取と次の撮影ポイントを探す作業に専念したようだ。

次郎助は一足早くベースキャンプに戻り、五右衛門風呂の準備をしていた。夕暮れ近く、ウィルソンは戻ってくると濡れた上着を脱ぎ、かじかんだ手をさすりながらほどよい温かさに焚かれた風呂の湯気を見て目を細めた。

新種の植物でも見つけたのだろうか、いつになく興奮した様子で機嫌がよかった。次郎助は不思議なほど、その時の状況を鮮明に覚えていた。

On February 19 and 20, Wilson took 18 photographs in total. On February 21, he took none. It may have been rainy, windy, or stormy with snow because snowfall is not unusual in this region at that time of the year. It seems that Wilson gave up on carrying camera equipment with him; instead, he focused on collecting plants and looking at new locations for taking photographs in the area.

Jirosuke had returned to the base camp a little earlier to prepare the bath. At sundown, Wilson came back, took off his wet jacket, and rubbed his hands together that were numb from the cold. His expression softened when he saw steam rising from the warm bath.

Wilson was extraordinarily excited and in a good mood. Perhaps he may have found a new species of plant. Strangely enough, Jirosuke clearly remembered this event.

Habitat of several trees. *Tsuga sieboldii* (Japanese hemlock), *Cryptomeria japonica* (Japanese cedar), *Stewartia monadelpha*, *Abies firma* (Japanese fir). ツガ、スギ、ヒメシャラ、モミの混成林。

Wilson Stump
ウィルソン株

通訳を介したウィルソンと森林官たちの会話が、次郎助の耳に入ってきた。
「あれは洞窟かもしれないね」
「まさか、あんなところに洞窟はないはずです。首を突っ込んだら中に引き込まれるかもしれませんよ」
「明日もう一度、あの場所へ行って調べてみよう」
ウィルソンは何か、とてつもないものと遭遇したのかもしれない。次郎助は、そんな予感で妙に胸が騒いだ。
翌日、身支度を整えると一行は安房川沿いをしばらく上流に上り、細く険しいけもの道のような小路を辿って標高1030mまで一気に登った。樹木の枝が幾重にも頭上で交差する下には、地衣類ですっぽりと包まれた森が続いていた。
柔らかな緑苔の敷物を踏んで、荘厳な建物の内部を探索するような神妙な心持ちで、ウィルソンはゆっくりと歩みを進めた。

Jirosuke overheard the conversation between Wilson and the foresters through the interpreter.
"That may be a cave."
"It can't be. There couldn't be a cave in that place. You might be dragged into it if you poked your head in there."
"Let's go to the site again tomorrow for an investigation."
Jirosuke had an odd premonition that Wilson may have encountered something extraordinary.
On the following day, the party members prepared themselves for an expedition. They walked up along the Anbo River for a while, and climbed up to one thousand thirty meters above sea level on a narrow and steep path which looked like an animal trail. Under thickly growing branches, the forest floor was fully covered with lichen. Wilson cautiously walked on the soft carpet of green moss with a solemn attitude, as if exploring the inside of a sacred building.

Cryptomeria japonica (Japanese cedar), girth 11ft.
Men on a plank walkway in the forest.
スギ　周囲 3.4m　森の中に敷設された丸太道を歩く男たち。

Wilson Stump
ウィルソン株

　朝霧が少しずつ夜の帳を引き上げていくにつれ、モノクロームの森に光の帯が降り注ぎ、その一条の光のなかに高さ4mほどのこんもりとした巨岩のような塊が見えた。ウィルソンが洞窟だと思ったのも頷ける。それはツタ類、苔類やたくさんの植物を身にまとい、旅人を誘い込むように南方にぽかっと大きな口を開けていた。24mほどの高さのスギが3本、護衛の兵士のように周りを取り囲んでいる。にわかには信じがたかったが、近寄ってよく見てみると、稀に見る巨大な屋久スギの切株に間違いなかった。

　森林官が切株の表面をびっしり覆っていた植物を切り払おうとしたその時、ウィルソンは厳しい口調でその動きを制した。「ダメだ。切らないで、そのままに」

　そして、植物を1種類ずつ丁寧に採取すると綿の古布に大事そうに包んだ後で、時間をかけて切株のサイズを測った。「胸高周囲50ft（15.2m）」とウィルソンはフィールド・ノートに数字を書き写すと撮影の準備に入った。

　As the morning mist slowly lifted the veil of night, the rays of sunlight started to penetrate into the monochrome forests. There appeared a gigantic mass that looked like a huge rock of about 13ft. high. Wilson reasonably considered this mass to be a cave. It was covered with ivy, moss, and many other plants, and was open to the south as if to lure tourists. Three Japanese cedars of about 80ft. tall surrounded this mass like bodyguards. When the party members closely observed the mass, they found to their surprise that it was a rare, massive Yakusugi stump.

　The moment a forester tried to cut away plants that thickly covered the surface of the stump, Wilson stopped him in a harsh tone. "No. Don't cut them. Leave as it is."

　Wilson carefully picked plants of one species and another, wrapped them carefully in old cotton cloth, and took time to measure the size of the stump. He recorded in his field notebook the circumference of 50 ft. (15.2 m) at chest height, and started to prepare for photographing.

Wilson Stump. Men standing by and upon the stump, 50ft. round, with three *Cryptomeria japonica* (Japanese cedar), height 80ft. × girth 11.5 and 10ft. growing thereon. alt.1030m
ウィルソン株　周囲 15.2m　上下左右に人物が配置されている。樹高 24m ×周囲 3.5m と 3m の 3 本のスギがその上に育っている。標高 1030m

Wilson Stump
ウィルソン株

カメラを慎重にセットした後で、ウィルソンは4人の男たちを上下左右にシンメトリーに配置した。左下の制服姿の男以外は、地元の若者たちが指名された。

株の左上で正面にまっすぐな視線を向けているのが次郎助、その右横で上体を少し斜めにしてはにかんだような表情をみせているのが喜平、そして右下の杉の根元に佇む大柄な青年が比賀之助であることが、それぞれの家族の証言により100年後の今、初めて明らかになった。

この一枚の写真は、ウィルソンが屋久島で撮ったほかの写真と比べてみると明らかに雰囲気が異なる。樹木の大きさを示すためのツールとして人を配置したというよりは、記念撮影的な意味合いが感じられる。このように4人もの人物をことさら大きく目立つように挿入した写真は、他にはないからだ。

なぜ、屋久島の旅のいわばクライマックスともいえる大株の写真をこのようなかたちで残したのだろうか。ウィルソンの特別な思い入れがそこにあったのてはないだろうか。

After setting up the camera with much care, Wilson arranged four men symmetrically. They were young local men, except the man in the uniform in lower left.

Jirosuke is at the upper left of the stump and looks straight at the camera. Kihei who is on Jirosuke's right looks bashful, with the upper part of his body slightly angled. Higanosuke is the large young man at the lower right, standing at the root of the stump. They were identified by their families for the first time, one hundred years after the photograph was taken.

The atmosphere of this photograph is obviously different from that of other photographs taken by Wilson on Yakushima. This one looks like a commemorative picture; it does not seem that human subjects were arranged just as tools for showing the size of the tree. No other photographs have as many individuals in such size and in such a conspicuous manner.

Why did Wilson take this type of photograph of the large stump, which could be considered the climax of his journey on Yakushima? He must have had a special reason.

Jirosuke Maki　Jan. 3rd,1890〜Jan. 31st, 1955.
(24 years old at that time)
牧次郎助　明治 23 年 1 月 3 日〜昭和 30 年
1 月 31 日　（当時 24 歳）

Kihei Oishi　Oct. 15, 1890〜Dec. 18, 1941.
(23 years old at that time)
大石喜平　明治 23 年 10 月 15 日〜昭和 16
年 12 月 18 日　（当時 23 歳）

Higanosuke Watanabe
Aug. 16, 1892〜Sep. 15, 1974.
(21 years old at that time)
渡辺比賀之助　明治 25 年 8 月 16 日
〜昭和 49 年 9 月 15 日　（当時 21 歳）

Wilson Stump
ウィルソン株

株のなかに足を踏み入れると、そこは大きな空洞になっていた。広さは31.4㎡、畳20枚ほどにもなる。屋久島の森で現存する最大のものだ。この大株は16世紀末に京都の方広寺を建立するために献上されたものではないかと言われている。伐採前の樹高は50mほどもある、想像を絶する巨木であったという。当時の残骸である先端部分や木片は、今なおそのまま近くに横たわっている。

株の付け根から清泉が湧き出ていて、内部には木製の祭壇と囲炉裏の跡があるのをウィルソンは見つけた。簡易の宿泊所として長く樵たちが使用してきたようだが、辺りの森に入ることができなくなってからは苔や小植物が着生するまま放置され、その存在が人々の記憶から忘れ去られようとしていたのだ。

ウィルソンが再発見して再びその巨大な姿を世にさらすようになって、この大株はいつしか山の聖地となった。株の入り口に村人の手で鳥居が建てられ、「大株神社」として敬われた。

現在は『ウィルソン株』と呼ばれ、屋久島のシンボル的な存在になっている。

They walked into the stump and found that the inside was a big cave measuring 338sq.ft. This is the largest existing stump in the forests of Yakushima. It is believed that the tree was cut at the end of the 16th century and offered for construction of a temple in Kyoto; the tree was about 164ft. tall, which is beyond our imagination. The top parts and fragments of the tree were found near the stump.

Clear spring water was flowing from the root of the stump. Inside a wooden altar and remnants of a fireplace were found. It seems that the stump had long been used by woodcutters as a simple shelter. After the forests around the stump became no longer accessible, the stump was left unattended, with moss and small parasitic plants growing on it. The presence of the stump was almost forgotten.

After rediscovery by Wilson and exposure to the public, this large stump became a sacred site of the island. A shrine gate was set up at the entrance of the stump. The stump was long worshipped by the local people as the large stump shrine. Today, it is called the "Wilson Stump" and serves as a symbol of Yakushima.

Trochodendron aralioides, height 55ft. × girth 23.5ft. crown 50ft. *Cryptomeria japonica* (Japanese cedar) to right. alt.750m

ヤマグルマ　樹高 17m ×周囲 7.2m　樹冠幅 15m　右にスギが見える。標高 750m

Endless Succession of Life
命のつながり

*I saw only on fallen trees and old stumps
the Cryptomeria successfully renewing itself unaided by man.*

倒れた木や切株の上でのみ、人の手を借りずにスギが更新しているのを私は見た。

　屋久スギは江戸時代の終わりまでにその70％が、主に年貢として屋根を葺く平木をとるために伐採された。山仕事に入った村人たちは、まっすぐな木材を得るためにスギの根元の広がった部分に櫓を組んでその上から幹を切り倒していた。そのため、森中に残された屋久スギの切株は3〜4mと背が高く、存在感があった。

　深山の森の床は倒れた木で埋め尽くされ、このような背の高い切株が至るところに点在し、ビロードのような苔がその上をびっしり覆っていた。

　気の遠くなるような長い年月を霧の深い厳しい自然の中で育った屋久スギは、倒れた後も息途絶えることはない。苔のマントの下では、残された切株は腐ることもなく数百年の歳月を見事に耐え、次の世代のためにひたすら命の炎を保ち続けているのだ。

By the end of the Edo period (1603–1867), 70% of Yakusugi were logged to produce roofing shingle materials that were primarily collected as a land tax. In the mountains, villagers set up scaffolding around the thick bottom part of the trees to cut the trunks at a certain height to procure straight-grained materials. That is why Yakusugi stumps left in the forests were tall, 10ft. to 13ft. high, and had unique presence.

The forest floor in the deep mountains was covered with fallen trees, and was studded with stumps that were thickly clad in velvet moss.

Surviving for an extensive period of time in the harsh, often fogbound environment, under a mantle of moss, the stumps are miraculously preserved for centuries without decaying. They persistently retain the spark of life for the next generation.

Cryptomeria japonica (Japanese cedar). Growing out of old stump, girth 12ft. with trees and vegetation growing upon them. Note rich Cryptogamic flora. alt.750m

さまざまな木や植物に覆われた大きなスギの切株。周囲 3.7m。豊かな苔の植生に注目。標高 750m

Endless Succession of Life
命のつながり

鬱蒼とした苔むした樹林のなかで、ウィルソンが見たものは、親スギの倒木や切株を栄養分として、新しく生まれたスギの種がその上に芽を吹き、すくすくと成長し命のバトンタッチを果たしている世代交代の姿だった。

「80年以上前に倒れた木と森林官は言うが、苔がはりついた外側5cmほど内側は、未だ完璧だった」

暗い林床では日光を好むスギの若木は生き残ることができないので、こうして親の世代が遺した体を借りてスギの種は命をつないでいくのだ。倒木や切株の上には、小さな苗木から高さ26mほどの大木までさまざまな成長過程のスギが見られた。

山の奥に入れば入るほど、そこには何千年という時の流れが始まりも終わりもないメビウスの輪のような時間軸のなかで、一瞬一瞬の時を刻み続けている。幾世代にもわたって脈々と更新されていく、生命の終わりなき連鎖の宇宙がそこにあった。

In the thickly moss-covered forests, Wilson witnessed the generational shift— new Japanese cedar seeds sprouting and growing rapidly using nutrients of fallen parent cedars and stumps.

"On such trees, which the foresters said had been felled eighty years before, the wood was still perfectly sound a couple of inches below the moss-clad exterior."

On the dark forest floor, saplings of Japanese cedar cannot survive due to lack of sunlight. Fallen trees and stumps provide nutrients to young trees to grow. In fact, trees in different life stages, ranging from small saplings to large trees of about 85ft. high, were observed.

Deeper in the mountains, time was passing like the Möbius strip — no beginning or end in cycles of thousands of years. The forests represented a microcosm of an endless chain of existence, passing life from one generation to the next.

Young trees, growing on old decayed stump of *Cryptomeria japonica* (Japanese cedar); the usual way of natural afforestation in the forests of Yakushima.

古いスギの切株の上に若いスギが育っている。屋久島の森でよく見られる自然植林のかたち。

Endless Succession of Life
命のつながり

生き抜くための巧みな戦略と努力を積み重ねているのはスギ族だけではない。倒木や切株の上でスギと仲良く一緒に生命を紡ぐヤマグルマの姿も多く見られた。
「この森で2本の木は、お互いに奇妙な偏愛を持っているに違いない」
スギとヤマグルマはふとしたきっかけで、倒れた幹や株の上で隣り合って芽を吹き、長い間仲良く一緒に育った。やがてスギがヤマグルマを追い越して高く空に聳え立つと、幼馴染みを風や嵐から守る役割を担うようになった。そんな微笑ましい情景が森のあちこちに見られた。しかし、たちの悪いヤマグルマもいて、成長するにつれてスギにしがみつき我が身だけが生き残ろうと必死の様相をみせていた。運悪くそのようなヤマグルマと隣り合わせになってしまったスギは、養分を吸い取られて今まさに絞め殺されようとしていた。
このような植物たちの生存競争の情景が、太古から続く長い映像のひとコマとなって目の前で演じられている様子には、ウィルソンも驚嘆するばかりだった。

Other plants were also using well-developed strategies to survive. Trochodendron were often seen growing in harmony with Japanese cedar on the fallen trees and stumps of Yakusugi.
"The two trees in these forests, it would seem, have a curious predilection for one another."
By chance, seeds of Japanese cedar and Trochodendron sprouted side by side on fallen trunks and stumps; the trees grew in harmony for many years. Before long, Japanese cedar rose higher, and served as shields to protect their "childhood friends" from winds and storms. Such friendly coexistence was seen everywhere in the forests. Some ill-natured Trochodendron clung to Japanese cedar as they grew older, desperately trying to survive at the expense of their partners. Unfortunate Japanese cedar were about to be strangled to death.
The struggle for survival of these plants constituted part of the long history of the forests from ancient times. The spectacle was simply awesome.

Cryptomeria japonica (Japanese cedar) and *Trochodendron aralioides*. Snow-covered mountain behind. alt.780m
スギとヤマグルマ。遠くに雪を頂いた山岳が見える。標高 780m

Endless Succession of Life
命のつながり

ウィルソンは、また別の面白い生命現象にも出くわした。2本か3本の異なる針葉樹が根元のところで互いにからみ合い、成長している姿をもカメラのレンズは捉えていた。

互いに接近しつつ両者ゆずらずに成長した結果、次第に寄り添うようになり、とうとう合体してしまい奇妙な姿態を見せる生物になったとみえる。

どちらがどちらに着生したのではなく、やむを得ない理由によりひとつになることで生きる道を選んだ例だ。

深い霧に覆われた山奥には幅広い世代のさまざま樹木が混然一体となって生き続け、それらにまた木ヅタなどのツタ科の植物やシダ、苔などの地衣類や菌類が着生し命を繋いでいた。まさに、沈黙の森のなかの壮大な生命のドラマだった。

中国の奥地でもこれほど豊かな植生の森はなかったと、ウィルソンは後日、屋久島の旅をそう振り返っている。

Wilson encountered another intriguing life phenomenon. His camera captured several conifers of different species entangled at their roots, coalescing into a single tree.

It seems that these trees grew closely without giving way, gradually closed the gap, and eventually coalesced. These trees chose to survive by unifying themselves, instead of being parasitic on one another.

In deep mountains shrouded in dense fog, trees of various ages and species were interdependent for survival. These trees served as hosts of vines as well as ferns and mosses.

A spectacular drama of life was performed in the tranquil forests. At a later date, Wilson recalled the journey in Yakushima, and wrote that forests of rich vegetation to this extent were never found elsewhere, even in inland China.

Cryptomeria japonica (Japanese cedar), girth 25ft. with young *Chamaecyparis obtuse* (Hinoki cypress) upon it. alt.800m

スギ、周囲 6m の上に若いヒノキが一体となって育っている。標高 800m

Legendary Pine
伝説のマツ

From sea-level to about 100m altitude is coastal savannah with scattered hamlets, patches of cultivation, isolated trees, chiefly Pinus Thunbergii Parl.

海岸から標高100mの地点は草原地帯で、点在する家屋や耕作地、そして主にクロマツがまばらに見られた。

旅の終わりが近づいていた。2月24日、ベースキャンプを後にして、来た道を辿りながらウィルソンは名残惜しげに森の姿をカメラに収め続けた。どのような風景に出会うか予測がつかなかった往路はガラス乾板を節約しながら登ってきたが、復路は気にかかる樹木に向かってシャッターを切りながら余裕をもっての下山となった。

やがて、楠川の村が真下に霞んで見えた。その向こうに青い海が広がり、漁船の白帆が波間に見え隠れするなか、海岸線のクロマツが地表に長い影を投げかけていた。

ウィルソンは夢から醒めた気持ちで、改めてここが小さな島であったことを思い出した自分に驚いた。室内から突然、明るい外に出た時のように唐突に森林が終わった先には、出発した時と変わらないのどかな里の風景があった。たった1週間ほどの旅であったが、それほど屋久島の森のなかの体験は別世界を旅していたような感覚に満ちていたのだった。

The end of the journey was drawing near. On February 24, Wilson continued to photograph the forests, feeling reluctant to leave. When he first arrived at the island and climbed the path, he saved glass plates because he did not know what landscape he would see. On his way back, he tripped the shutter without economizing glass plates to photograph trees in which he had taken an interest.

At the end of stone steps, Kusugawa was faintly visible right below the party. Beyond the village spread the blue ocean, with white sails of fishing boats appearing in the waves. Trees of Black Pine cast long shadows on the coastline.

Wilson felt as if he awoke from a dream; he was surprised to remember that he was on a small island. The forests came to an abrupt end; it was like going outdoors in the daytime from inside a building. There was the same peaceful village landscape that he saw when he departed. Given that the journey was only about one week, the forests on Yakushima offered an extraordinary experience of traveling in a totally different world.

Curving *Clethra barbinervis*, height 35ft. × girth 5ft. *Abies firma* (Japanese fir) to left. alt. 800m

曲がったリョウブ　樹高 11m ×周囲 1.5m　左にモミ。標高 800m

Legendary Pine
伝説のマツ

楠川の集落へ下りきった所で、魔法の国への案内人を務めてくれた若者たちとの別れが待っていた。山のなかでは逞しく見えた彼らは早くも里の顔に戻り、ウィルソンと固い握手を交わすと集落の中へ静かに溶け込んでいった。

ウィルソンはその日は宮之浦の宿に戻り、翌日にかけて収集した植物の整理とガラス乾板の梱包に費やした。そして26日、舟を使って海岸線を北上して近隣の村々を探索した。

ひとかたまりになった家々が点在する海岸の里では漁業が盛んに営まれていた。単体で立つクロマツの大木は、それぞれの集落で名前を付けて呼ばれ、鳥居が横に据えられて祭りや祈りの場所とされていた。集落と山の間に帯状に広がる草原にもクロマツの姿はまばらにあったが、海辺で見られるような、ずば抜けて大きな木は見られなかった。

これらの木は、あとどれくらい長く里の風景を彩り続けられるのだろうか。マツのように風によって花粉を受粉する樹木は、集団としてある程度まとまった密度が必要だった。

When the party reached Kusugawa, Wilson had to bid farewell to the young men who had served as guides to the wonderland. These young men, who appeared so robust in the mountains, already looked like ordinary villagers. After a firm handshake with Wilson, they quietly returned to village life.

On that day, Wilson returned to the Miyanoura inn, and spent time organizing his collected plants and packaging glass plates until the end of the following day. On February 26, he headed north in a boat along the coast to explore nearby villages.

In coastal villages, housing clusters were sparsely located. Fishery was a thriving business. Large single Black Pine were nicknamed in respective villages. Shrine gates were built next to them. The sites were used for festivals and prayers. Black Pine were also scattered in the belt of grassland between villages and the mountains, but extraordinarily large trees like the ones seen on the coast were not found.

How long would these trees survive as part of the village landscape? Trees including pines need to grow in clusters because they are pollinated by wind.

Pinus thunbergii (Japanese black pine). The coastal savannah, sea level to 150m.
海岸から標高 150m 地点に広がる草原地帯に見えるクロマツ。

Legendary Pine
伝説のマツ

　水辺には、バナナとヘゴなどに混じってさまざまな木が生えていた。海岸の植物でウィルソンが興味を持って採取したのはヤマモモ、トベラ、シギ、アセビ、アブラギリなどだった。

　実はウィルソンには、この時どうしても気になるマツがあった。屋久島に渡る前に種子島で多く見たゴヨウマツだった。森林官の話では、屋久島の西部の山にも同じものがたくさん生息しているという。それは、中国本土で採取したタカネゴヨウと非常に似通っていながら、細かいところに微妙な差が見られた。

　中国のマツが遠い昔に日本の南の島にたどり着き、そこで固有種を形成したものではないか。屋久島旅行に同行した鹿児島大林区署の三好哲男はその後、両島で採取した球果をウィルソンに送り、ウィルソンはそれを基に帰国してから論文のなかで自論を展開している。短い時間で、その類似性を見逃さなかったプラント・ハンターの観察力はさすがである。

　ウィルソンの推測通りそのマツは後日、日本人植物学者によって「ヤクタネゴヨウ」と命名され、屋久島、種子島の固有種として正式に登録された。

The shore was home to various species of trees, in addition to banana and large spiny tree fern.

There was a pine that strongly attracted Wilson's attention. It was similar to Pinus *Armandi franch* (chinese white pine); he observed the trees in large numbers on Tanegashima before visiting Yakushima. According to foresters, mountains in the western part of Yakushima were home to these trees. The cones of this pine tree were very similar to, but subtly different in detail, from those he collected in Mainland China.

Later, Tetsuo Miyoshi of the Kagoshima Forestry Bureau, who accompanied Wilson in his journey on Yakushima, sent pinecones collected on these islands to Wilson. After returning home, Wilson asserted in his paper that, in ancient times, pinecones in China were carried to Japan's southern islands where endemic species were formed. As a plant hunter, he was a keen observer in identifying the similarities of plants in a short space of time.

As Wilson conjectured, the pine was later named "Yakutanegoyo" by a Japanese botanist, and was officially registered as an endemic species of Yakushima and Tanegashima.

Pinus thunbergii (Japanese black pine), height 65ft. × girth 9ft. village of Shitogo behind.
クロマツ　樹高 20m ×周囲 2.7m　志戸子の集落が後ろに見える。

Legendary Pine
伝説のマツ

　ウィルソンのカメラが100年前に捉えたクロマツの大木は、現在そのほとんどが屋久島の海岸から姿を消している。原因は過度の伐採や害虫被害、環境の変化などによって個体数が激減し、ウィルソンが危惧したように、世代交代が困難になっていったためだ。

　屋久島の海岸を色濃く縁どっていたマツの木陰は、トビウオの産卵場所でもあった。5〜6月、トビウオが産卵の時期を迎えると、マツの下はトビウオの卵で真っ白になった。村人はそこを「マツのソノ」と呼び、大量のトビウオを網ですくって手に入れていた。マツの姿が消えていくとともにトビウオ漁も舟で沖合に繰り出す形に変わり、魚の数は急減していった。

　丸木舟や柱の材料として珍重されたヤクタネゴヨウは、いまは環境省のレッドデータリストで「絶滅危惧種」に指定され、その存続が危ぶまれている。

　Today, most of the large Black Pines photographed by Wilson a century ago are gone from the coasts of Yakushima. Due to excessive logging, damage from harmful insects, and environmental changes, among others, the population of Black Pines decreased gradually, which made it difficult for the trees to breed.

　The shade of pine trees that rimmed the coasts of Yakushima served as spawning grounds for flying fish. In May and June, which was the spawning season, the areas under the pine trees turned snow-white with their eggs. Villagers called these areas "matsu-no-sono"; they were able to scoop a vast amount of fish with nets along the shore. As pine trees disappeared, villagers were forced to go offshore in their boats. The catch of flying fish decreased sharply.

　Black Pine, which was highly valued as the material for dugout canoes and pillars, is now an endangered species as designated in the Red List issued by the Ministry of the Environment, Government of Japan.

Ficus pumila. Clothing three Cherry trees. *Piper futo-kadsura*. With torii(shrine gate), near village of Shitogo. alt. 30m

オオイタビと3本の桜の木にまとわりつくフウトウカズラ。志戸子集落の近くの鳥居のそば。標高 30m

Legendary Pine
伝説のマツ

The more simple the people, the greater their appreciation.

人はシンプルであればあるほど、感謝の思いも大きい。

ウィルソンが宮之浦の港を後にしたのは、上陸してから10日後、2月27日のことだった。そして、二度と屋久島の地を踏むことはなかった。しかし、彼の心の中では屋久島の将来に対する深い憂いがあった。

「日本人の学者は屋久島のことのほとんどふれておらず、その植生についての研究はあまりされていないように思う」

鹿児島に着くとすぐに、ウィルソンはボストンへの手紙のなかでその思いを綴っている。屋久島の植物について科学的な調査が未だきちんとされていないのが気がかりだった。

ウィルソンはその後、サハリンに至るまで日本国中を旅して、1915年1月帰国の途に就いた。どこに行っても「あまりに近代化されすぎていて」、どこへでもついて回る営林署職員を有難いとは思いながらも「いつも行動を注意深く観察されている」と感じたウィルソンであった。そんな彼にとって、近代化に染まることなく止むことなき生活の喜びにあふれていた屋久島の佇まいと、その山のなかで地元の若者たちと過ごした時間は、日本旅行のなかでも特に忘れえぬ思い出として心に残ったに違いない。

一方、屋久島では『ウィルソン株』の名とともに、ウィルソンの姿は人々の記憶に深く刻まれた。

Wilson left the port of Miyanoura on February 27, 10 days after landing. He never came back to Yakushima. However, he was concerned about its future.

"I do not think Japanese botanists have given much study to the flora of this outlying group." As soon as he arrived in Kagoshima, Wilson wrote this in a letter to Boston. He was anxious because scientific studies had not been properly conducted about the plants of Yakushima.

Later, Wilson traveled across Japan up to Sakhalin, and left Japan to return home in January 1915. Wherever he visited, Wilson felt that the rest of Japan was influenced "too much by so-called civilization," and that "my movements have been carefully watched" because officers of the Forestry Bureau always accompanied him.

In his trip in Japan, Wilson particularly cherished the memories of Yakushima, the island where daily life was filled with ever-lasting pleasures away from the influence of modernization and he spent time with local young men in the mountains.

On Yakushima, Wilson's visit remains deeply engraved in people's memory thanks to the name of "Wilson Stump."

Pittosporum tobira. Shrub spontaneous habitat by edge of sea. Bush 8ft. Head 12ft.
一湊の海岸に自生するトベラの茂み　樹高 2.4m ×樹冠幅 3.7m。

When you destroy the forests you upset the balance of nature. The present generation must remember that it only holds the forests in trust for future generations.
Ernest Henry Wilson

あなたが森を破壊する時、あなたはまた自然のバランスを壊しているのだ。
将来の世代のために今の世代ができることは、森を守っていくことに尽きるということを忘れてはいけない。
アーネスト・ヘンリー・ウィルソン

81

100年後への伝言

　1年間の日本滞在中、ウィルソンが移動した総距離は実に15000kmに及ぶ。全部で600枚を超える写真のガラス乾板と200種の植物標本は、100年前の日本の植物相や自然環境を知る貴重な記録となった。帰国後、ウィルソンは『日本の桜』と『日本の針葉樹』の2本の論文を書いて出版した。特に後者の論文のなかで、屋久島の植生についての記述に多くのページを費やし、初めて屋久島の名を世界に紹介した。

　2年後の1917年（大正6年）2月、ウィルソンは再び横浜の港に下り立った。琉球、伊豆大島、小笠原、韓国、台湾への調査が未着手だったからである。最初の訪問地の琉球に向かう途中、鹿児島高等農林学校（現・鹿児島大学）で田代善太郎と偶然に出会った。そして帰りにわざわざ鹿児島市郊外の加治木町まで足を延ばし、当時そこで中学の教員をしていた田代の自宅を訪ねた。同行したのは、屋久島を一緒に旅した三好哲男だった。この訪問は個人的なものであったのだろう、ウィルソンが残した手紙などの記録のなかでは触れられていない。

　一方、田代は日記にこの日のことを書き残していた。要約すると、「3月16日午前11時48分の列車にてハーバード大学のウィルソン氏が来る。自宅に案内して食事をしたのち海岸まで散歩し、5時56分に見送る」とある。つまり、ウィルソンと田代は6時間ほど一緒に過ごした。そこでどういう会話が交わされたのか詳細は書かれていないが、直後から田代は屋久島の植物の資料を求め始めているのである。明らかに、ウィルソンの示唆がそこにあったと思われる。

　翌年、田代は屋久島に渡りウィルソンが言及した大株を確認、株の中で一夜を明かした。さらに教職を辞すると内務省の天然記念物調査嘱託を自ら希望し、屋久島の調査担当の辞令を受けた。そして、1923年（大正12年）までに合計5回屋久島の植物調査を行い、『屋久島天然記念物調査報告書』を書き上げた。この時初めて、屋久島の森の本格的かつ科学的な調査が行われた。屋久杉原生林が国の天然記念物に指定されたのは、その翌年のことだった。

　その後、戦後の高度成長期に大量伐採の波が屋久島の森に襲いかかるが、島民、学者、自然愛好家らを中心とした自然保護運動の機運の高まりとともに、国立公園に編入されるなど保護への道筋がつけられていった。

　100年前にウィルソンが撮った里の風景と現在のそれを比べてみると、山の稜線は変わらないばかりか、禿げ山だった里山は豊かな緑に覆われている。また、山中の苔に覆われた森林は現代の登山者を引きつけてやまない。

　1993年12月、屋久島は日本で初めてユネスコの世界自然遺産地に登録された。ウィルソンが発した百年前の警鐘が、その最初の一歩となったといっても過言ではない。

Photograph taken in Kagoshima the day after Wilson was visiting Mr. Zentarou Tashiro, with Tetso Miyoshi(right) and Tomoji Ushio (back), March 17, 1917
ウィルソンが田代善太郎の自宅を訪れた翌日に鹿児島で撮影されたもの。三好哲男（右）と牛尾朋次（後）1917年3月17日

田代善太郎著『鹿児島県屋久島の天然記念物調査報告（復刻版）』

Heritage from 100 Years Ago

Wilson's stay in Japan lasted for a year and his exploration covered a total distance of fifteen thousand kilometers. He shipped back over six hundred glass plate negatives and samples of two hundred plant specimens, which have become precious records for studying flora and the natural environment of Japan a hundred years ago. After his return to the US, Wilson wrote and published two papers on Japan "The Cherries of Japan" and "The Conifers and Taxads of Japan." He spent many pages on descriptions of vegetation on Yakushima in the latter, introducing the name of the island to the world for the first time.

In mid-February 1917, Wilson landed at the Port of Yokohama for the second time. His purpose was to explore the Ryukyu Islands, Izu Oshima Islands, Bonin Islands, Korea, and Formosa. On his way to the Ryukyu Islands he became acquainted with Zentaro Tashiro in Kagoshima. After his expedition in Ryukyu, he extended his stay in Kagoshima for one day to visit Tashiro's home in Kagikicho, where he was a junior high school teacher. There is no trace of this call in any of Wilson's official documents. It was a completely private visit that he had made of his own accord.

"The Conifers and Taxads of Japan" E.H.Wilson, 1916.

On the other hand, Tashiro wrote about that day in his diary. In summary he wrote, "March 16. Met Mr. Wilson of Harvard University at the station at 11:48 am. Brought him home, enjoyed lunch, and took a walk to the beach. Saw him off at 5:56." This means that Wilson and Tashiro spent about six hours together. Tashiro does not go into the details of their talks, but immediately after the visit he started collecting references on plants in Yakushima and appealed to the government to conduct an investigation to designate Yakushima's primeval forests as a natural monument. It is evident that Wilson had advised him to do so. We cannot be mistaken in assuming that Wilson entrusted the official protection of Yakushima to Tashiro.

The following year, Tashiro sailed to Yakushima for the first time, confirmed the presence of the great stump that Wilson had mentioned, and spent the night in the hollow of that stump. He went on to resign from teaching, volunteered to work part-time for the Ministry of Home Affairs on the investigation and was appointed to do so. By 1923, he conducted a total of five vegetation surveys on Yakushima and compiled "A report on the investigation to designate Yakushima as a natural monument." These became the first full-scale and scientific vegetation surveys of Yakushima's forests. The primeval forests were designated as a natural monument the following year.

A wave of massive lumbering with chainsaws destroyed some woods of Yakushima in Japan's postwar development period, but a stop was put to this in 1954 with the establishment of the primeval Yakusugi forests as a Special Monument Area. Islanders and scholars took the lead in movements to protect the natural environment, and Yakushima was specified as a National Park Special Area in 1964, paving the way for its conservation.

In December 1993, Yakushima became the first Japanese site to be designated as a UNESCO World Natural Heritage Site. It is no exaggeration to say that the warnings Wilson voiced one hundred years ago triggered the first step toward this designation.

Sequel of the Three Young Men 若者たち、その後

牧次郎助

　ウィルソンが「ジロさん」と呼んで可愛がった次郎助は、ブラジルに移住する夢を持っていたが、「山を守れ」というウィルソンの言葉がその思いを引き止めた。多くの島の男たちがそうであったように、屋久スギの板木を背負って下ろす作業に従事しながら、漁船を所有しトビウオ漁でも生計を立て、65歳で没するまで島の外に出ることもなく穏やかな生涯を送った。

　ひとり息子の市助が高校生になった時、次郎助は初めて若き日にひとりの西洋人と山で過ごした思い出を語って聞かせた。そして、貧しい暮らしのなかから学資をねん出して息子を鹿児島の大学に送り出した。市助は教員となり、島の小学校で15年間教鞭をとった後、県内各地の小学校で教職に勤しんだ。退職後は屋久島に戻って教育長に任命され後輩の指導に励んだ。この間、数えきれないほどの多くの教え子たちに、自然の素晴らしさ、屋久島の山の大切さを教えた。次郎助は自分の息子を学問の道に進ませることで、ウィルソンとの約束を果たしたのだ。

　焼酎が入ると、次郎助は屋久島の島唄をよく唄った。「屋久のお嶽をおろかに思うなよ、金の蔵よりなお宝」（まつばんだ）。それは、ウィルソンが次郎助に語った言葉そのものだった。

Jirosuke Maki

　Fondly called "Jiro-san" by Wilson, Jirosuke originally had dreams of immigrating to Brazil. However, he decided not to do so at Wilson's words, "Protect the mountains of Yakushima." Like many men of the island, he never left it, leading a simple life carrying Yaku cedar boards down from the mountains and catching flying fish on his fishing boat untill he passed away at the age of sixty-five.

　When his only son Ichisuke became a high school student, Jirosuke told him of his memories of a foreigner he spent time with in the mountains in his youth. He managed to save money from his poor income to send his son to study at a university in Kagoshima. Ichisuke became a teacher. He taught at an elementary school on the island for fifteen years, then at elementary schools throughout Kagoshima Prefecture. He returned to the island after retirement and was appointed superintendent of education of Yakushima and instructed younger teachers. Throughout his career he taught countless students about nature's greatness and about how invaluable the mountains of Yakushima are. Jirosuke kept his promise with Wilson by giving his son a good education.

　Jirosuke loved to sing when he drank shochu. His favorite was an old folksong of Yakushima, "Matsubanba."

　It goes "Never think lightly of the mountains of Yaku, more precious are they than treasures of gold."

　That is the exact message that Wilson gave to Jirosuke.

Sequel of the Three Young Men 若者たち、その後

大石喜平

「腰が痛くてね、整理しきれんのよ。よかったら見ていって」

喜平の息子、鴻の妻の大石トシ子は大きな木の箱を指差した。その中には、夫の遺品であるさまざまなノート、画集が乱雑に詰められていた。それは、父親の喜平から聞いた屋久島の昔の暮らしの断片を鴻が詩や文章やイラストのかたちで書き残したものだった。もちろん、そのなかにはウィルソンとの思い出に触れたものもあった。

大石家は代々、指物師だった。喜平もまた、物静かな職人気質をもった人であった。山や海の仕事に携わりながら、喜平は木の弁当箱、煙草盆、道具入れなど美しい民具をたくさん作った。

そして、独特な表現で自分の体験を息子に伝えた。それは、五感で経験したことを形に残す道であった。

「思いやりがあれば、花を見ることができる。これから何億年先の人たちが花と交流できるか、これは人間の心ひとつにある」

鴻の遺品のなかに散りばめられている言葉の奥には喜平の姿があり、その向こうには間違いなくウィルソンの背中があった。

父親とともに山で多くの時を過ごした鴻は生命の素晴らしさを自然から学び、言葉を紡ぎ絵を描く行為を通して、島の歴史を記録し保存することに専念して一生を終えた。

Kihei Oishi

"I've been unable to organize these since my back hurts, but please feel free to take a look at them," said Toshiko Oishi, wife of Hiroshi who was Kihei's son, as she pointed to a big wooden box. The box was full of notes and sketchbooks that her husband left behind when he passed away. Hiroshi made poems, stories, and illustrations of the old life in Yakushima that he had heard from his father, Kihei. Needless to say, among them were memories of Wilson.

Kihei's family were joiners for generations. Kihei was adept at handwork and made many beautiful wood lunchboxes, tobacco trays, toolboxes and other articles for daily use between working in the mountains and at sea. A quiet craftsman, Kihei told his son of his own experiences in a unique poetic way. He passed on his experiences through his gifted five senses.

Hiroshi wrote, "If you are thoughtful, you can see flowers. Whether people living hundreds of millions of years from now can communicate with flowers depends on how considerate people are."

Behind the words left in the works of Hiroshi, is Kihei, and beyond Kihei is Wilson.

Hiroshi spent much of his time helping his father in the mountains and learned the wonders of life from nature. His life was one of a person preserving the history of the island in the forms of literature and art.

Sequel of the Three Young Men 若者たち、その後

渡辺比賀之助

　人並み外れた力持ちでウィルソンを驚かせた比賀之助の武勇伝は、今も村の語り草になっている。100kgほどの生きた豚を背負い、山奥の集落に売りに行ってその日のうちに帰ってきたこと。戦時中、疎開で家を山中に移す際に大きな柱を背負ってひとりでそっくり建て替えたこと。医者ぎらいで、山仕事中にナタで足を大怪我した時も自分で薬草を傷口に詰め込み、熱した油を注ぎ込んで治したこと。大男でいかつい顔をしていたが、心根は優しかった。しかし、あまりに声が大きいので、村の幼い子供は比賀之助を見ると泣き出した。

　比賀之助は「純粋な山男」だった、と息子嫁の渡辺チエ子は振り返る。夜明け前に家を出て山に入り、日暮れとともに帰ってきた。妻を早く亡くしてからは自分のスギ山に小屋を建て、多くの時をそこで過ごした。83歳で亡くなる直前まで、木に登って枝切りをしていた。

　ひとり息子の浩は会社勤めだったが父の亡くなった後に山小屋に行き、その丁寧で心のこもった仕事ぶりに深く感銘を受けた。退職後は父の跡を継いで山仕事に関わって生涯を終えた。

　比賀之助は言葉としては何も残さなかったが、自らの生きざまを通して山の仕事の大切さを息子に伝えた。無口で頑固者だった比賀之助ならではのやり方でウィルソンの遺産を継承したのだった。

Higanosuke Watanabe

　Higanosuke amazed Wilson by climbing steep slopes with a heavy iron cauldron which served as a bathtub on his back. Many a tale of his recklessness is still told in the village today. Once he carried a live swine as heavy as one hundred kilograms on his back to a hamlet deep in the mountains to sell and came back the same day. When houses were evacuated and moved into the mountains during the war, he carried big pillars on his back, and rebuilt his house all by himself. He hated doctors, and when he injured his foot seriously with a hatchet doing mountain work, he stuffed mugwort into the cut, poured heated oil onto it, and healed it by himself. He was a big man with a stern look, but was kind at heart. Yet, small children would start crying when they saw him because he had such a loud voice. Many more tales remain.

　Discribed by his daughter in law, Chieko Watanabe, as a true man of the mountains, Higanosuke dedicated his whole life to mountain work. He left for the mountains before dawn and came home at sunset. After the untimely death of his wife, he built a hut in his cedar mountain. He climbed trees and cut branches until the time just before he died at the age of eighty-three.

　His only son, Hiroshi, was a company worker and visited the mountain hut after his father's death. He was so deeply impressed with the careful and mindful work of his father, that after retirement, Hiroshi succeeded his father in forest work. Higanosuke did not leave any words behind, but he passed down the significance of mountain work through his own way of life. Higanosuke too, passed on Wilson's heritage in his own wordless and stubborn way.

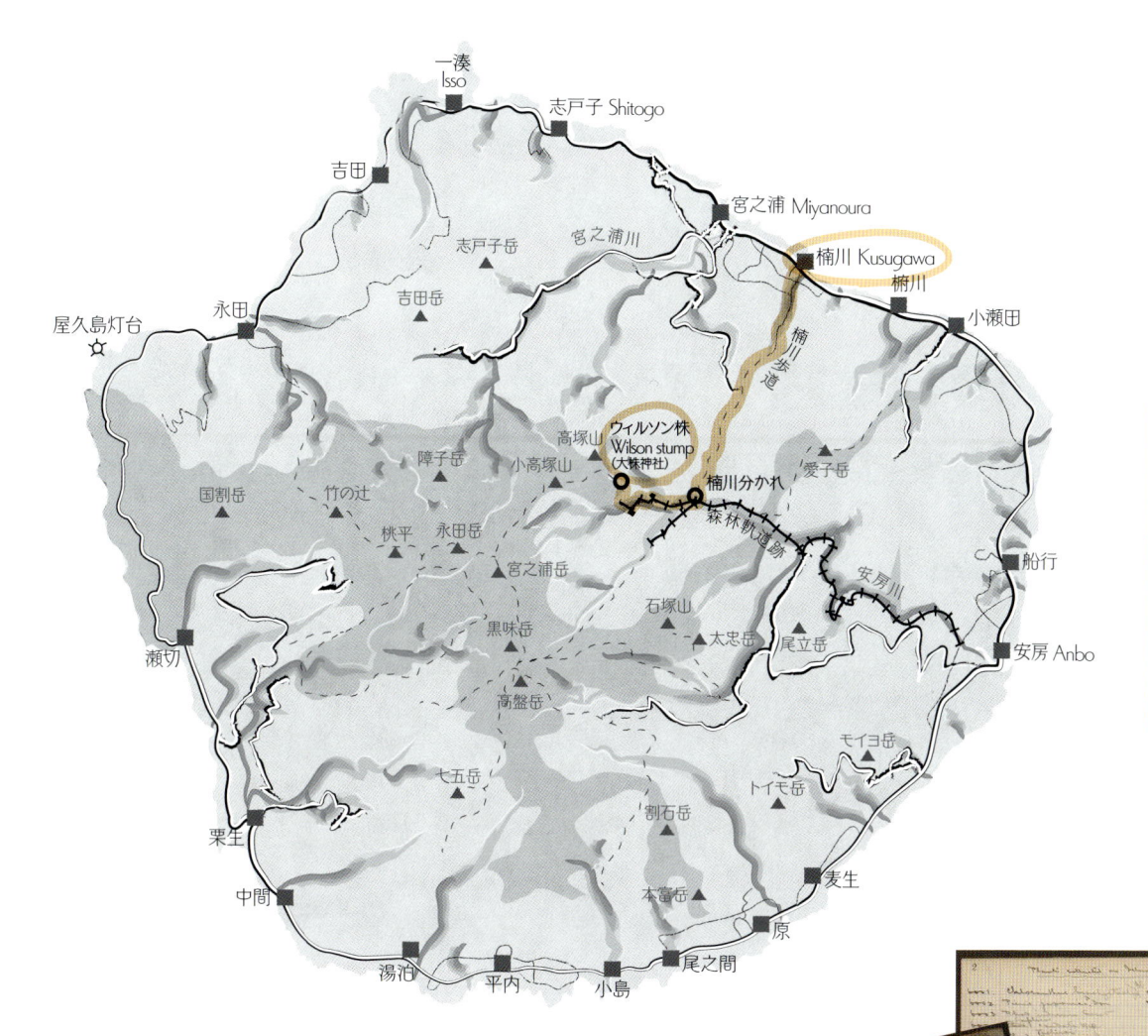

Wilson Trail
ウィルソンの歩いた道

凡　例	
	世界遺産登録区域
	川
▲	山
	自動車道
	一般道路
- - -	登山道
┼┼┼┼	森林軌道跡

『屋久島地図 2013 年』
'Map of Yakushima, 2013'.

Wilson's handwritten field notebook.(Feb.1914-Jan.1915) You can see the word "Yakushima Feb.18th 1914" on the upper right of the first page.

ウイルソン手書きのフィールド・ノート（1914 年 2 月～1915 年 1 月）。最初のページ右上に「屋久島 1914 年 2 月 18 日」の記述が読める。

ウィルソンの生涯

　アーネスト・ヘンリー・ウィルソンは、1876年2月15日イギリス中西部の小さな村、チッピング・カムデンで6人兄弟の長子として生まれた。家計を助けるため、小学校卒業と同時に13歳から地元の園芸店で働き始めたのが、植物に興味を持つきっかけだった。
　16歳の時にバーミンガム植物園の庭師見習いとして迎えられ、そこで働きながら技術学校の夜学部に通い植物学の基礎を身につけていった。5年後にはその優秀な成績が認められ、首都ロンドン市内にあるキュー王立植物園に21歳で雇用された。

Wilson(right) with dogs in China, 1908.
中国探検中のウィルソン（右）。1908年

当時のウィルソンの目標は教員資格を取得して故郷で安定した収入を得る職に就き、幼い兄弟姉妹を養っていくことだった。
　そんな折、人生の転機がやってきた。ロンドンの有名な種苗会社ヴィーチ商会が中国にプラント・ハンターを送る計画を立て、相談を受けたキュー王立植物園は若干22歳のウィルソンに白羽の矢を立てたのだった。プラント・ハンターは、未だ知られざる植物を発見し採集するのを目的に未踏の地に派遣される探検家のことで、その任務は肉体的にも精神的にも強靭さを必要とされるばかりでなく、身の危険を伴う仕事でもあった。
　1899年、ウィルソンはボストン経由で香港に入り3年間にわたる中国探検（1899〜1902）に出発した。チベットとの国境地帯で'幻の花'と言われたハンカチノキをはじめ約400の新種の植物を発見するなど、この旅で大きな成果を収め、その後プラント・ハンターとしてのキャリアの階段を一気に上りつめていくことになる。
　帰国後、婚約者と教会で結ばれた半年後には、2回目の中国探検（1903〜1906）に旅立った。彼が渡された指示書には、やはりチベット国境にだけ咲くイエローポピーの種を確保することだった。数々の身の危険を冒した2万kmの旅を経て、標高3500mの高山でウィルソンは目的の植物を見つけ、ヴィーチ商会から41個のダイヤモンドで装飾された金のピンを贈られた。
　1906年には娘のミリエルが誕生したが、愛娘の存在もウィルソンンをフィールドから引退させることはできなかった。第

3回中国探検（1907〜1909）、第4回中国探検（1910〜1911）はアメリカのハーバード大学アーノルド植物園の要請によるもので、前の2回と異なり樹木の科学的な調査を主とした旅だった。休む暇なく立て続けに中国奥地を訪れ、ウィルソンはことごとく偉業を達成した。

中国への最後の旅となる4回目の探検は、大きな代償を伴うものだった。突然の土砂崩れのためウィルソンは右脚に大怪我を負ったのだ。医者もいない奥地での出来事だった。しかし、ウィルソンはその手に旅の大きな目的物のひとつのリーガル・リリーの球根をしっかり握りしめていた。この美しいユリは、ウィルソンが見つけた数多くの新種希少植物のなかでも代表的な花となった。

アメリカのメディアはこの生死をかけた偉業を褒め称え、ウィルソンはいつしか「チャイニーズ・ウィルソン」の名で呼ばれるようになり、英米の植物学会などから数多くの賞が授与された。

ウィルソンの残りの人生は、採取した植物の整理と著述、そして研究旅行に費やされた。1914年には日本での針葉樹とサクラの調査に着手し、南は屋久島から北はサハリンまで約1年かけて日本縦断旅行を敢行した。1917年には韓国、台湾などにも足を伸ばし、1920年から1922年にかけてはオーストラリア、ニュージーランド、タスマニア、インド、ケニア、南アフリカなど世界の庭園を回った。

未知の地域に足を踏み入れ集めたのは1000以上の新種の植

Family portrait in Tokyo, 1914.
ウィルソン一家のポートレート。東京 1914年

物、16000の標本、そして訪れた土地を記録した2600枚にのぼる写真だった。23年間にわたる探検の旅を基にウィルソンは多くの論文や専門書を著すだけでなく、さまざまな園芸雑誌に一般読者向けの読み物を載せ、各地で講演を行い、ラジオショーにも出演するなど最も人気のある「伝説のプラント・ハ

ンター」として名声を馳せた。
　日本、特に屋久島では「ウィルソン博士」と呼ばれ、名門ハーバード大学所属の学究肌の教授といった堅いイメージを持たれているが、ウィルソンがハーバードの学位（M.A.）を授けられたのは、日本の旅から帰った翌年だった。信じがたいことだが、この時までウィルソンはアカデミックな資格を持っていなかった。20代初めから旅の人生を送ってきた彼には、大学で単位をとる機会を得る時間がどこにもなかったからだ。しかし、短い滞在の間に屋久島、種子島の固有種ヤクタネゴヨウの存在を看破したことにみられるように、植物に関する観察力、記憶力においては人並みはずれた才能を持っていた人であった。
　学者というよりは基本的には行動する人であり、フィールドワークが彼の本領だったといえる。自然と接すること、そのために肉体的なチャレンジを楽しむことに何よりも幸せを感じた。それゆえに、森の生態系保存の必要性や地球の将来に対する危惧についても、その著書や講演で発信し続けた人でもあった。
　「時にして友人が言う。君は地球を歩き回ってさぞかし困難に耐えたことだろうと。確かにそうだが、そんなことは私にとって何でもない。なぜなら、私は自然の広大な場所に住み、その喜びを深く飲み干したのだから」
　1927年からはアーノルド植物園園長に就任し、家族と共にニューイングランド近隣で小旅行に出掛け静かな時間を楽しんだ。そして3年後、ボストン郊外で自身が運転していた車が崖

Mrs.Wilson and Muriel in the tea ceremony in Tokyo,1914.
着物姿のウィルソン夫人と娘のミリエル。茶の湯に興じている風情。東京1914年

から転落し、助手席に乗っていた妻とともにこの世に別れを告げた。皮肉にも事故の原因は、道路に積もった枯葉で車輪がスリップしたことによるものだった。
　20年以上住んだにもかかわらず、ウィルソンはアメリカの市民権はとらなかった。最後までイギリス人であり続けることにこだわった彼の遺志に基づき、ウィルソンの遺骨はイギリス領カナダのモントリオールの霊園に眠っている。
　54歳の短い生涯だったが、常にフィールドを活躍の場とし、精力的に植物と関わり続けた輝かしくも鮮烈なナチュラリストの人生であった。

The Life of Ernest Henry Wilson

Ernest Henry Wilson was born at Chipping Campden, a small village in mid-western England, on February 15, 1876, as the eldest of six children. On finishing elementary school, he started working at a local nursery where he became interested in plants.

At the age of sixteen, he was employed at the Birmingham Botanical Gardens as an apprentice gardener and in the evening after work, he studied the basics of botany at a technical school. Having earned excellent grades, he was hired at the Royal Botanical Garden at Kew in London five years later at the age of twenty-one. He hoped to obtain a teacher's license and earn a stable income in his hometown so he could support his younger brothers and sisters.

Then came the turning point in his life. James Veitch and Sons, a nursery firm which had planned to send a plant hunter to China, asked the Kew Gardens to recommend someone, and Wilson was chosen. Plant hunters were explorers sent to distant lands to discover and collect unknown plants. It was a dangerous mission requiring physical and mental strength.

In 1899, Wilson entered Hong Kong by way of Boston and set off on an expedition to China which lasted for three years. On this expedition he discovered approximately four hundred new plant species including the legendary *Davidia involucrata*(Dove tree) which he found near the border of Tibet. With this grand success, his career rapidly took off.

He married his fiancée upon his return to England, and departed for his second Chinese expedition (1903-1906) a half year later. This time, he was directed to secure the seeds of *Meconopsis integrifolia*(yellow Chinese poppy) which bloom only near the Tibetan border. In his adventures full of perils covering twenty thousand kilometers, Wilson found the plant he was after in a peak three thousand five hundred meters above sea level. Veitch rewarded him with a gold pin adorned with forty-one diamonds for his achievements.

Even the birth of his sweet daughter Muriel in 1906 could not keep him away from the fields. He continued to explore China, on his third (1907-1909) and fourth (1910-1911) expeditions. These differed from

Mr. and Mrs. Wilson and Muriel with three unidentified women in Japan, 1914.
ウィルソン一家と3人の女性たち。1914 年

the former two in that he was to conduct scientific investigations of trees at the request of the Arnold Arboretum of Harvard University. He was constantly exploring the wilderness of China, and with noteworthy outcomes each time.

He was to pay heavily on his fourth and last expedition to China. His leg was severely injured in a sudden landslide. The accident occurred in the outback where no doctors were to be found. Nonetheless, his hand was tightly holding bulbs of *Lilium regale*(Regal lily) which was one of his pursuits of that journey. This beautiful lily was to become a flower representing the numerous rare plants that he discovered.

The American media praised him for his great achievements at the risk of his life, and called him "Chinese Wilson." He earned many awards from American and British botanical societies.

The remainder of Wilson's life was spent organizing the plant specimens he had collected, writing books, and taking research trips. He began investigating plants in Japan in February 1914, covering the whole archipelago from the southern island of Yakushima to Sakhalin in the north taking a whole year. Three years later in 1917, he investigated plants in Korea, Formosa (now Taiwan) and a few other places. From 1920 to 1922 he visited gardens around the world including Australia, New Zealand, Tasmania, India, Kenya, and South Africa.

His acquisitions from unknown lands amasses over one thousand new plant species, sixteen thousand samples, and two thousand six hundred photographs that he took in those lands. Based on his journeys over twenty-three years, Wilson wrote stories for general readers in gardening magazines as well as numerous academic papers and books. He also gave countless lectures and performed on radio shows at various places, making him famous as a legendary plant hunter.

In Japan, and especially in Yakushima, people called him Doctor Wilson, giving him the image of a serious academic professor of the prestigious Harvard University. However it was only the year after he returned from Japan that he received a Master of Arts degree from Harvard. Believe it or not, Wilson did not have any academic licenses before then. He had neither time nor opportunity to acquire credits at universities, as he had always been traveling since his early twenties.

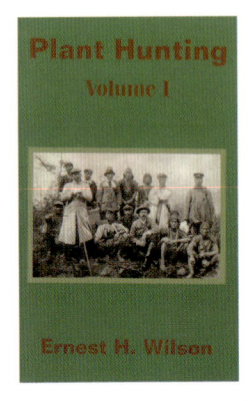

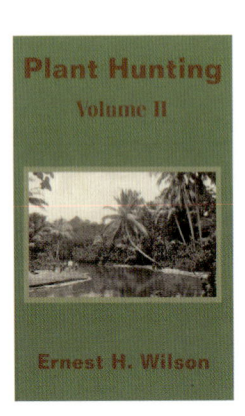

Ernest Henry Wilson (1927) *"Plant Hunting Volume* Ⅰ & Ⅱ*"*
アーネスト・ヘンリー・ウィルソン著（1927）
『プラント・ハンティング Ⅰ & Ⅱ』

But we must say that he had remarkable talents in observing plants and remembering them, as is proven by distinguishing that *Pinus amamiana*(known as Yakutanegoyo in Japanese) is an endemic pine tree of Yakushima and Tanegashima islands during his short visit.

He was basically a person of actions, rather than a scholar, and fieldwork suited his true nature. He found utmost pleasure in encountering nature and meeting physical challenges to that end. This was why he persistently sent out messages on the necessity to preserve the ecological system of the forests for the future of our planet through his books and lectures.

"Sometimes friends have said 'You must have endured much hardship wandering in out of the way corners of the earth.' I have. But such count for nothing, since I have lived in Nature's boundless halls and drank deeply of her pleasures."

In 1927, Wilson became the Keeper of the Arnold Arboretum, and enjoyed taking small peaceful trips around New England with his family. Three years later, he was killed in a car accident with his wife. His car fell off a cliff when he was driving in the suburbs of Boston. Ironically, the tires had slipped on fallen leaves on the road.

Though Wilson lived in the States for more than twenty years, he did not acquire US citizenship. Respecting his will to stay British, he was buried on British territory in a graveyard in Montreal, Canada.

In his short life of fifty-four years, Wilson lived in the fields vigorously engaging with plants as a glorious and extraordinary naturalist.

Wilson standing at the front entrance of the Hunnewell administration building at the Arnold Arboretum, 1922.
アーノルド植物園管理棟前の玄関に立つウィルソン。　1922年

Acknowledgments
謝辞

感謝を込めて（敬称略）
牧市助・大石トシ子・渡辺チエ子・牧良平・鎌田道隆・松田賢志
その他、楠川集落の方々
・
ウィリアム・フリードマン（ハーバード大学アーノルド植物園園長）・
リサ・ピアソン（アーノルド植物園図書館館長）

In Appreciation
Ichisuke Maki, Toshiko Oishi, Chieko Watanabe
Ryouhei Maki, Michitaka Kamata, Kenshi Matsuda
and many of Kusugawa Village People
・
William Friedman, Director of the Arnold Arboretum of Harvard University.
Lisa E.Pearson, Head of the Library and Archive of the Arnold Arboretum

植物名アドバイス
寺岡行雄（鹿児島大学農学部生物環境学科教授）
前田三文（林野庁九州森林管理局屋久島森林生態系保全センター所長）

Plant Adviser
Yukio Teraoka, Prof.of Kagoshima University
Mifumi Maeda, Director of Yakushima Forest Ecosystem Conservation Center

装丁デザイン＆英訳アドバイス
ウィリアム・ブラワー

Book Design & English Translation Adviser
William Brouwer

協力
林野庁九州森林管理局屋久島森林生態系保全センター
公益財団法人屋久島環境文化財団
屋久島町

Cooperation
Kyusyu Regional Forest Administration Office, Yakushima Forest Ecosystem Conservation Center
Yakushima Environmental Culture Foundation
Yakushima Town

著者
古居 智子（ふるい ともこ）
大阪生まれ。北海道大学卒。米国ボストンでジャーナリストとして活躍後、1994年米国人建築家の夫とともに屋久島に移住。NPO法人屋久島エコ・フェスタ理事長。環境保護活動に励みながら、島の暮らし、歴史、自然をテーマに執筆活動を続けている。

著書：『夢見る旅「赤毛のアン」』（文藝春秋）『屋久島　恋泊日記』（南日本新聞社）『屋久島　島・ひと・昔語り』（南日本開発センター）『密行　最後の伴天連シドッティ』（新人物往来社）『はじまりのかたち―屋久島民具ものがたり』（NPO法人屋久島エコ・フェスタ）

Author
Tomoko Furui

Born in Osaka. Graduated from Hokkaido University. Worked as a journalist in Boston, U.S.A. Moved to Yakushima with her American husband in 1994. Serves as Director of NPO Yakushima Eco-festa. Committed to writing books about life, history, and nature of Yakushima while working on activities to protect the environment.

References
参考文献

〈日本文資料〉 Japanese Sources

田代善太郎『鹿児島県屋久島の天然記念物調査報告（復刻版）』1995 年 ㈲生命の島

田代晃二編著『田代善太郎日記　大正編』1972 年　創元社

山本秀雄編著『屋久島歴史小年表』2007 年　㈲生命の島

太田五雄編著・発行『屋久島山岳体系（五高山岳部復刻版）』2013 年

牧良平『屋久島物語』1996 年　マキノンブル社

岩川興助『わが回想記』1962 年　非売品　岩川先生銅像建立発起人会

楠川区しゃくなげ会編『ちょっぽい 1 － 15 号』1993 － 2001 年　非売品

山本秀雄『文献資料紹介≪第 4 回≫　アーネスト・ヘンリー・ウイルソン「日本の針葉樹」』　生命の島第 4 号　㈲生命の島

金谷整一『屋久島自然系　その 1　絶滅の危機にあるヤクタネゴヨウ』生命の島第 52 号　㈲生命の島

上屋久町郷土誌編集委員会『上屋久町郷土誌　全一巻』1984 年

屋久町郷土誌編さん委員会『屋久町郷土誌　第三巻　村落誌下』2003 年

杉崎元『ウイルソン株』1956 年　熊本営林局林友　暖友林　7 月号

金平亮三『ウイルソン君を懐ふ』1931 年　台湾博物学会報第 21 巻 113 号　台湾博物学会

F. キングドン - ウォード『植物巡礼ープラント・ハンターの回想』1999 年　塚谷裕一訳　岩波文庫

アリス・M・コーツ『プラントハンター東洋を駆けるー日本と中国に植物を求めてー』2007 年　遠山茂樹訳　八坂書房

ロバート・フォーチュン『幕末日本探訪記　江戸と北京』1997 年　三宅馨訳　講談社学術文庫

白幡洋三郎『プラントハンター』2005 年　講談社学術文庫

〈英文資料〉 English Sources

E. H. Wilson (1927). *Plant Hunting, Volume Ⅰ & Ⅱ*. University Press of the Pacific Honolulu, Hawaii.

E. H. Wilson (1913). *A Naturalist in Western China 1913*. Cambridge Library Collection, Cambridge University Press.

E. H. Wilson (1916). *The Conifers and Taxads of Japan*. Cambridge Printed at The University Press.

E. H. Wilson (1917). *Aristocrats of the Garden*. Doubleday, Page & Company.

Roy W Briggs (1993). *'Chinese' Wilson A Life of Ernest H Wilson 1876-1930*. The Royal Botanic Gardens, Kew The Royal Botanic Garden, Edinburgh.

James Herbert Veitch(1906). *Hortus Veitchii A history of the Rise and Progress of the Nurseries of Messrs. James Veitch and Sons*. James Veitch & Sons Limited, Chelsea.

Edward I. Farrington(1931). *Ernest H. Wilson Plant Hunter*. The Stratford Company.

Peter J. Chwany. *E.H.Wilson, Photographer*. A.A.A. of Harvard University, Boston.

Alfred Rehder. *Ernest Henry Wilson*. A.A.A. of Harvard University, Boston.

Richard A. Howard. *E.H.Wilson as a Botanist*. A.A.A. of Harvard University, Boston.

E. H. Wilson. *Correspondence 1899-1930. from Japan, Feb. 1914-Jan. 1915*. A.A.A. of Harvard University, Boston.

E. H. Wilson. *Field Notes on collected plants and seed* feb.1914-Jan.1915. A.A.A.of Harvard University, Boston.

Gwen Bell. *E.H."Chinese" Wilson, Plant Hunter*. Seattle Washington Journal American Rhododendron Society.

※ A.A.A. = Archives of the Arnold Arboretum

Wilson's Yakushima Memories of the Past
ウィルソンの屋久島 100年の記憶の旅路

2013年11月23日　初版第一刷　発行

著者	古居智子
写真提供	ハーバード大学
	日本カメラ博物館
装丁・本文デザイン	佐藤遥子
進行	堀野恵子
発行人	前田哲次
編集人	谷口博文
発行所	KTC中央出版
	東京都台東区蔵前 2-14-14
	〒111-0051
	電話　03-6699-1064
	ファックス　03-6699-1070
印刷・製本	株式会社廣済堂

内容に関するお問い合わせ、ご注文などはすべて上記KTC中央出版までお願いします。乱丁、落丁本はお取り替えいたします。本書の内容を無断で複製・複写・放送・データ配信などすることは、かたくお断りいたします。
定価はカバー表示してあります。

ISBN978-4-87758-370-5　C0072　©Tomoko Frui, Printed in Japan